Hua Xu

Multi-Modal Sentiment Analysis

Hua Xu (iD)
Department of Computer Science and Technology
Tsinghua University
Beijing, P. R. China

ISBN 978-981-99-5775-0   ISBN 978-981-99-5776-7   (eBook)
https://doi.org/10.1007/978-981-99-5776-7

© Tsinghua University Press, Beijing, P.R.China 2023
Jointly published with Tsinghua University Press, Beijing, P.R.China
The print edition is not for sale in Mainland China. Customers from Mainland China please order the print
book from: Tsinghua University Press.

This Springer imprint is published by the registered company Springer Nature Singapore Pte Ltd.
The registered company address is: 152 Beach Road, #21-01/04 Gateway East, Singapore 189721,
Singapore

Paper in this product is recyclable.

Preface

The natural interaction ability between human and machine mainly involves human-machine dialog ability, multi-modal sentiment analysis ability, human-machine cooperation ability, and so on. In order to realize the multi-modal sentiment analysis ability of intelligent computer, it is necessary to make the computer own strong multi-modal sentiment analysis ability in the process of human-computer interaction. This is one of the key technologies to realize efficient and intelligent human-computer interaction.

The research and practical application of multi-modal sentiment analysis is oriented to human-computer natural interaction; this book mainly discusses the following levels of hot research contents: Multi-modal Information Feature Representation, Feature Fusion and Sentiment Classification. Multi-modal sentiment analysis oriented to natural interaction is a comprehensive research field involving the integration of natural language processing, computer vision, machine learning, pattern recognition, algorithm, robot intelligent system, human-computer interaction, etc. In recent years, our research team from State Key Laboratory of Intelligent Technology and Systems, Department of Computer Science, Tsinghua University, has conducted a lot of pioneering research and applied work, which have been carried out in the field of multi-modal sentiment analysis for natural interaction, especially in the field of sentiment feature representation, feature fusion, robust sentiment analysis based on deep learning model. Related achievements have also been published in the top academic international conferences in the field of artificial intelligence in recent years, such as *ACL, AAAI, ACM MM, COLING*, and well-known international journals, such as *Pattern Recognition, Knowledge based Systems, IEEE Intelligent Systems*, and *Expert Systems with Applications*. In order to systematically present the latest achievements in multi-modal sentiment analysis in academia in recent years, the relevant work achievements are systematically sorted out and presented to readers in the form of a complete systematic discussion.

Currently, the research on multi-modal sentiment analysis in natural interaction develops fast. The author's research team will timely sort out and summarize the latest achievements and share them with readers in the form of a series of books currently. This book can not only be used as a professional textbook in the fields of natural interaction, intelligent question answering (customer service), natural language processing, human-computer interaction, etc., but also as an important reference book for the research and development of systems and products in intelligent robots, natural language processing, human-computer interaction, etc.

As the natural interaction is a new and rapidly developing research field, limited by the author's knowledge and cognitive scope, mistakes and shortcomings in the book are inevitable. We sincerely hope that you can give us valuable comments and suggestions for our book. Please contact xuhua@tsinghua.edu.cn or a third party in the open-source system platform https://github.com/thuiar/Books to give us a message. All of the related source codes and datasets for this book have also been shared on the following websites https://github.com/thuiar/Books .

The research work and writing of this book were supported by the National Natural Science Foundation of China (Project No. 62173195), National Science and Technology Major Project towards the New Generation of Broadband Wireless Mobile Communication Networks of Jiangxi Province (03 and 5G Major Project of Jiangxi Province, Grant No: 20232ABC03402) and High-level Scientific and Technological Innovation Talents "Double Hundred Plan" of Nanchang City in 2022. We deeply appreciate the following students from State Key laboratory of Intelligent Technology and Systems, Department of Computer Science, Tsinghua University for their hard preparing work: Xiaofei Chen and Jiayu Huang. We also deeply appreciate the following students for the related research directions of cooperative innovation work: Zhongwu Zhai, Wenmeng Yu, Kaicheng Yang, Jiyun Zou, Ziqi Yuan, Huisheng Mao, Wei Li, and Baozheng Zhang. Without the efforts of the members of our team, the book could not be presented in a structured form in front of every reader.

Beijing, China   Hua Xu
June 2023

Contents

1 Overview 1
  1.1 Overview of Multimodal Sentiment Analysis 1
      1.1.1 Overview of Research on Multimodal Sentiment
             Analysis 2
      1.1.2 Overview of Related Research on Modality Loss 5
      1.1.3 Conclusion 6
  1.2 Overview of Multimodal Machine Learning 7
      1.2.1 Overview of Multimodal Representation Learning 7
      1.2.2 Overview of Multimodal Representation Fusion 10
      1.2.3 Conclusion 12
  1.3 Overview of Multitask Learning Mechanisms 12
      1.3.1 Multitasking Architecture in Computer Vision 13
      1.3.2 Multitasking Architecture in Natural Language
             Processing 14
      1.3.3 Multitasking Architecture in Multimodal Learning 15
      1.3.4 Conclusion 18
  1.4 Summary 18
  References 19

2 Multimodal Sentiment Analysis Data Sets and Preprocessing 23
  2.1 Multimodal Sentiment Analysis Datasets 23
      2.1.1 Introduction 23
      2.1.2 CMU-MOSI 24
      2.1.3 CMU-MOSEI 25
      2.1.4 IEMOCAP 25
      2.1.5 MELD 25
      2.1.6 Conclusion 25
  2.2 Multimodal Sentiment Analysis Dataset with Multilabel 26
      2.2.1 Introduction 26
      2.2.2 CH-SIMS Dataset 28

	2.2.3	Multimodal Multitask Learning Framework	31
	2.2.4	Experiments	33
	2.2.5	Conclusion	36
2.3		An Extension and Enhancement of the CH-SIMS Dataset	38
	2.3.1	Introduction	38
	2.3.2	CH-SIMS V2.0 Dataset	40
	2.3.3	Feature Extraction	43
	2.3.4	Acoustic Visual Mixup Consistent (AV-MC) Framework	44
	2.3.5	Experiments	44
	2.3.6	Conclusion	49
2.4		Summary	49
References			50

3 Early Unimodal Sentiment Analysis of Comment Text Based on Traditional Machine Learning 53

3.1		Identifying Evaluative Sentences in Online Discussions	54
	3.1.1	Introduction	54
	3.1.2	The Proposed Technique	54
	3.1.3	Experiments	62
	3.1.4	Conclusion	67
3.2		Grouping Product Features Using Semisupervised Learning with Soft-Constraints	67
	3.2.1	Introduction	67
	3.2.2	The Proposed Algorithm	68
	3.2.3	Generating SL Using Constraints	71
	3.2.4	Distributional Context Extraction	72
	3.2.5	Experiments	73
	3.2.6	Conclusion	81
3.3		Constrained LDA for Grouping Product Features in Opinion Mining	82
	3.3.1	Introduction	82
	3.3.2	The Proposed Algorithm	82
	3.3.3	Constraint Extraction	87
	3.3.4	Experiments	88
	3.3.5	Conclusion	94
3.4		Product Feature Grouping for Opinion Mining	96
	3.4.1	Introduction	96
	3.4.2	The Proposed Soft-Constrained Algorithm	96
	3.4.3	Extracting the Example Set Using Constraints	98
	3.4.4	Distributional Context Extraction	100
	3.4.5	Experiments	101
	3.4.6	Conclusion	104

3.5 Exploiting Effective Features for Chinese Sentiment
      Classification 105
      3.5.1 Introduction 105
      3.5.2 Methodology 105
      3.5.3 Experimental Setup 110
      3.5.4 Experimental Results 111
      3.5.5 Conclusion 114
3.6 An Empirical Study of Unsupervised Sentiment Classification
      of Chinese Reviews 115
      3.6.1 Introduction 115
      3.6.2 Proposed Technique 115
      3.6.3 Empirical Evaluation 118
      3.6.4 Conclusion 122
3.7 Feature Subsumption for Sentiment Classification in Multiple
      Languages 123
      3.7.1 Introduction 123
      3.7.2 The Proposed Algorithm 123
      3.7.3 Experimental Setup 126
      3.7.4 Experiments 127
      3.7.5 Conclusion 131
3.8 Summary 132
References 133

4 Unimodal Sentiment Analysis 135
4.1 Text Sentiment Analysis Based on Word2vec and SVM^{perf} 135
      4.1.1 Introduction 135
      4.1.2 Methodology 136
      4.1.3 Experiments 140
      4.1.4 Conclusion 146
4.2 Contextual Heterogeneous Feature Fusion Framework
      for Audio Sentiment Analysis 147
      4.2.1 Introduction 147
      4.2.2 Proposed Method 147
      4.2.3 Experiments 152
      4.2.4 Conclusion 155
      4.2.5 Introduction 155
      4.2.6 Methodology 156
      4.2.7 Experiments 164
      4.2.8 Conclusion 172
4.3 Summary 174
References 174

5 Cross-Modal Sentiment Analysis 179
5.1 The Acoustic Visual Mixup Consistent (AV-MC) Framework 179
      5.1.1 Introduction 179

5.1.2   Multimodal Sentiment Analysis (MSA) Background   180
5.1.3   Automatic Sentiment Computing Approach with Modality
            Mixup Strategy   182
5.1.4   Experiments   189
5.1.5   Conclusion   195
5.2   Cross-Modal Sentiment Recognition Based on Hierarchical
         Grained and Acoustic Features   195
5.2.1   Introduction   195
5.2.2   Problem Definition   196
5.2.3   Methodology   197
5.2.4   Experiments   199
5.2.5   Conclusion   202
5.3   Cross-Modal Sentiment Classification for Alignment
         Sequences   202
5.3.1   Introduction   202
5.3.2   Methodology   203
5.3.3   Experiments   207
5.3.4   Conclusion   213
5.4   Summary   213
References   214

6   **Multimodal Sentiment Analysis**   217
6.1   Multimodal Sentiment Analysis Model Based
         on Self-Supervised Multitask Learning   217
6.1.1   Introduction   217
6.1.2   Methodology   218
6.1.3   Experiments   224
6.1.4   Conclusion   228
6.2   Multimodal Sentiment Analysis Method Based on Modality
         Missing   228
6.2.1   Introduction   228
6.2.2   Methodology   229
6.2.3   Experiments   233
6.2.4   Conclusion   238
6.3   Summary   239
References   239

7   **Multimodal Sentiment Analysis Platform and Application**   241
7.1   An Integrated Platform for Multimodal Sentiment Analysis   241
7.1.1   Introduction   241
7.1.2   Platform Architecture   242
7.1.3   Experiments   246
7.1.4   Model Analysis Demonstration   249
7.1.5   Conclusion   252

7.2 Robust Multimodal Sentiment Analysis Platform 252
    7.2.1 Introduction 252
    7.2.2 Demonstrating Robust-MSA 252
    7.2.3 Engaging the Audience 256
    7.2.4 Conclusion 256
7.3 Summary 257
References 257

Appendix 259
Symbol Cross-Reference Table 259
Code Link Table 261

About the Author

Hua Xu is a leading expert on Intelligent Natural Interaction and Service Robots. He is currently a Tenured Associate Professor at Tsinghua University, Editor-in-Chief of *Intelligent Systems with Applications* and Associate Editor of *Expert Systems with Application*. Prof. Xu has authored the books *Data Mining: Methodology and Applications* (2014), *Data Mining: Methods and Applications-Application Cases* (2017), *Evolutionary Machine Learning* (2021), Data Mining: Methodology and Applications (2nd edition) (2022), *Natural Interaction for Tri-Co Robots* (1) *Human-machine Dialogue Intention Understanding* (2022), and *Natural Interaction for Tri-Co Robots* (2) *Sentiment Analysis of Multimodal Interaction Information* (2023), and published more than 140 papers in top-tier international journals and conferences. He is Core Expert of the No.03 National Science and Technology Major Project of the Ministry of Industry and Information Technology of China, Senior Member of CCF, member of CAAI and ACM, Vice Chairman of Tsinghua Collaborative Innovation Alliance of Robotics and Industry, and recipient of numerous awards, including the 2nd Prize of National Award for Progress in Science and Technology, 1st Prize for Technological Invention of CFLP, 1st Prize for Science and Technology Progress of CFLP, etc.

List of Figures

Fig. 1.1   Example of multimodal data 2
Fig. 1.2   A typical multimodal sentiment analysis framework 4
Fig. 1.3   Nonalignment characteristics of text, audio, and video modal
            sequence features 5
Fig. 1.4   Factors that may trigger modal random deficits 6
Fig. 1.5   The architecture of TCDCN 13
Fig. 1.6   Cross-Stitch network structure. Each task has a separate
            network, but the Cross-Stitch unit linearly combines
            information from parallel layers of different task networks 14
Fig. 1.7   Cascade information of various tasks supervised in each layer.
            Lower-level tasks are supervised in the earlier layers 16
Fig. 1.8   The proposed OmniNet architecture [57] 17
Fig. 2.1   An example of the annotation difference between CH-SIMS
            and other datasets 27
Fig. 2.2   The distribution of sentiment over the entire dataset in four
            annotations 29
Fig. 2.3   Multimodal multitask learning framework 31
Fig. 2.4   Visualization in Unimodal Representations 37
Fig. 2.5   Illustration of the constructed CH-SIMS v2.0 dataset 39
Fig. 2.6   Statistics of each component of the CH-SIMS v2.0 dataset 41
Fig. 2.7   Left: the distribution of sentiment over the entire dataset in one
            Multimodal annotation and three unimodal (Text, Acoustic,
            and Visual) annotations. Right: the confusion matrix shows the
            annotations difference between different modalities in CH-SIMS
            (s) v2.0. The larger the value, the greater the difference 42
Fig. 2.8   Acoustic Visual Mixup Consistent (AV-MC) Framework under
            semisupervised learning paradigm, which consists of Multitask
            Late Fusion Backbone (**a**) with Modality Mixup Module (**b**) 45
Fig. 2.9   Case study results on CH-SIMS v2.0 48
Fig. 3.1   Overview of the proposed technique 55
Fig. 3.2   Interaction modeling of aspects, evaluation words, and emotion
            words (IAEE) 58

Fig. 3.3   Influence of λ on IAEE 66
Fig. 3.4   Influence of ρ on IAEE 66
Fig. 3.5   Influence of context window size 80
Fig. 3.6   Computing the weights for must-topics and cannot-topics 85
Fig. 3.7   RI results of constrained-LDA and the original LDA 91
Fig. 3.8   Comparisons with mLSA 92
Fig. 3.9 λ's influence on the overall performance 94
Fig. 3.10 η's influence on the overall performance (#topics $= 40$) 95
Fig. 3.11   Influence of the context window size on SC-EM 105
Fig. 3.12   The framework of sentiment classification based
              on supervised learning 106
Fig. 3.13   Generalized suffix tree constructed by the two short
              sentences: "I like this camera" and "because this camera looks
              like chocolate." The dotted arrows represent suffix links 108
Fig. 3.14   The comparison results on hotel data set 113
Fig. 3.15   The comparison results on product data set 113
Fig. 3.16   Domain-dependent characteristics of SNW. (**a**) Product review,
              (**b**) Hotel review 121
Fig. 3.17   Influence of the lexicons' scale 121
Fig. 3.18   The framework of the proposed algorithm 123
Fig. 3.19   The experiment on the mixed-language dataset
              (DF + "tfidf-c") 128
Fig. 3.20   Left: the accuracies achieved by the proposed algorithm
              on three open sentiment datasets in different languages. Right:
              comparisons of transductive learning (green and red) and
              inductive learning (dark blue) 129
Fig. 4.1   The general framework of our work 136
Fig. 4.2   The performance of lexicon-based with various C 145
Fig. 4.3   The performance of part-of-speech-based with various C 145
Fig. 4.4   Overall architecture of the proposed heterogeneous feature
              fusion framework 148
Fig. 4.5   The structure of the residual block with 128 filters 149
Fig. 4.6   The structure of the Contextual Single Feature Model 151
Fig. 4.7   The minimum and maximum accuracy of different models
              for audio sentiment analysis on the MOSI dataset 154
Fig. 4.8   Coattentive multitask convolutional neural network 158
Fig. 4.9   T-SNE visualization of the outputs in the final hidden layer
              of the BaseDCNN and our approach 173
Fig. 5.1   Multimodal sentiment analysis backbone 181
Fig. 5.2   Automatic sentiment computing approach with modality
              mixup strategy 183
Fig. 5.3   Modality Mixup strategy 184

Fig. 5.4   Semisupervised experiments with varying quantities of supervised
            data, the horizontal axis represents the quantity of supervised data,
            the vertical axis represents MAE performance, and different
            colours represent the quantity of fixed and different unsupervised
            data. Fixed the quantity of unsupervised data, changed the
            quantity of supervised data, and conducted semisupervised
            experiments 194
Fig. 5.5   Overall architecture of the proposed HGFM framework 198
Fig. 5.6   The confusion matrix of different hierarchical features in
            emotional categories. (a) Single feature. (b) Fusion feature 201
Fig. 5.7   Overview architecture of the Cross-Modal BERT Network 204
Fig. 5.8   The architecture of the masked multimodal attention 205
Fig. 5.9   Visualization of the attention matrices 212
Fig. 6.1   The overall architecture of Self-MM 219
Fig. 6.2   Unimodal label generation example 221
Fig. 6.3   The overall framework of TFR-Net which contains three modules:
            feature extraction module, modality reconstruction module, and
            modality fusion module 230
Fig. 6.4   Metrics curves of various missing rates on MOSI dataset 236
Fig. 6.5   Metrics curves of various missing rates on SIMS dataset 237
Fig. 7.1   The overall framework of the M-SENA platform contains four
            main modules: data management module, feature extraction
            module, model training module and model evaluation module ... 243
Fig. 7.2   Intermediate result analysis for TFN model trained on MOSI
            dataset 250
Fig. 7.3   On-the-fly instance test example 251
Fig. 7.4   Noise influence demonstration. (a): Noise injection,
            (b): feature and prediction comparison 254

List of Tables

Table 2.1   Summary of public multimodal emotional analysis dataset   24
Table 2.2   Baseline dataset settings   24
Table 2.3   Statistics of SIMS dataset   29
Table 2.4   Dataset splits in SIMS   34
Table 2.5   (%) Results for sentiment analysis on the CH-SIMS dataset   35
Table 2.6   (%) Results for unimodal sentiment analysis on the
             CH-SIMS dataset using MLF-DNN   38
Table 2.7   Multimodal sentiment analysis model on CH-SIMS v2.0
             dataset   46
Table 3.1   Summary of the four datasets   62
Table 3.2   Comparison results   64
Table 3.3   Data sets and gold standards   74
Table 3.4   Comparison results (L = 30% of the gold standard data)   78
Table 3.5   Influence of the seeds' proportion (which reflects the size
             of the labeled set L)   79
Table 3.6   Summary of the data sets   89
Table 3.7   Number of the extracted constraints   90
Table 3.8   Datasets, feature expressions, and synonym groups   101
Table 3.9   Performance of feature expression classification and
             clustering methods on various domains (best performance in
             each category in bold)   102
Table 3.10   Performance of feature expression classification and
             clustering methods with varying percentages of labeled
             seeds   103
Table 3.11   The criteria for key-node selection   109
Table 3.12   The summary of the data sets   110
Table 3.13   Performance of the N-gram-based features on two data sets   112
Table 3.14   Performance of the substring-based features on two
             data sets   112
Table 3.15   The summary of all sentiment lexicons   118
Table 3.16   Summary of the datasets   118

Table 3.17 Experimental results based on the original sentiment lexicons
            OL (without SNW removal) 119
Table 3.18 Improvements made by SNW removal 120
Table 3.19 The summary of the open datasets 126
Table 3.20 Comparisons with the best performance of the existing
            typical methods 127
Table 4.1   The key parameters of training command 137
Table 4.2   A brief summary of data sets 141
Table 4.3   The results of similar features clustering based on word2vec ... 143
Table 4.4   The performance of sentiment classification based
            on lexicon 144
Table 4.5   The performance of sentiment classification based
            on part-of-speech 144
Table 4.6   Accuracy for different combinations with two feature
            selection methods 144
Table 4.7   The experimental results for audio sentiment analysis on the
            MOSI dataset. M: spectrum features. S: statistical features 153
Table 4.8   The experimental results for audio sentiment analysis on the
            MOUD dataset. M: spectrum features. S: statistical features 153
Table 4.9   BaseDCNN architecture for single tasks, FLD, and FER 156
Table 4.10 Facial expression and facial landmarks annotations statistics
            of RAF, SFEW2, CK+, and Oulu-CASIA. "NA" means
            vacancy 163
Table 4.11 (%) Results for FER on RAF and SFEW2 with different
            multitask learning methods "BaseDCNN (FER)" and
            "BaseDCNN (FLD)" are the single-task baselines for FER and
            FLD tasks, respectively 165
Table 4.12 (%) Results for FER on CK+ and Oulu-CASIA with different
            multitask learning methods "BaseDCNN (FER)" and
            "BaseDCNN (FLD)" are the single-task baselines for FER and
            FLD tasks, respectively 166
Table 4.13 (%) Comparisons with state-of-the-arts on the RAF database ... 168
Table 4.14 (%) Comparisons with state-of-the-arts on the SFEW2
            database 169
Table 4.15 (%) Comparisons with state-of-the-arts on the
            CK+ database 170
Table 4.16 (%) Comparisons with state-of-the-arts on the Oulu-CASIA
            database 170
Table 4.17 (%) Transfer validation on different train sets and test sets 171
Table 4.18 Time cost comparison between our method and baselines
            on the RAF database during the inference stage. Batch $= 32$ ... 174
Table 5.1   Model performances for single task Multimodal Sentiment
            Analysis model on CH-SIMS dataset. The best results are
            highlighted in bold 191

Table 5.2   Model performances for Traditional Multimodal Sentiment
             Analysis model on CH-SIMS v2.0 dataset. Models with
             (∗) are trained on multitasking. The best results are highlighted
             in bold 192
Table 5.3   Model performances for unimodal sentiment analysis with
             modality mixup strategy. Models with (†) are trained with
             modality mixup strategy 193
Table 5.4   Model performances for cross-modality sentiment analysis with
             modality mixup strategy. Models with (†) are trained with
             modality mixup strategy. Models with (∗) are trained on
             multitasking 193
Table 5.5   Model performance for Semisupervised Sentiment Analysis.
             Models with (∗) are trained on multitasking, (s∗ represents the
             additional use of unsupervised data) 193
Table 5.6   Statistics of utterances for IEMOCAP and MELD datasets 199
Table 5.7   The overall performance on IEMOCAP and MELD datasets
             comparison with the state-of-the-art 200
Table 5.8   Performance of each emotional category 201
Table 5.9   Experimental results on CMU-MOSI dataset. The best results
             are highlighted in bold. h means higher is better and means
             lower is better. T text, A audio, V video 211
Table 6.1   Dataset statistics in MOSI, MOSEI, and SIMS 225
Table 6.2   Results on MOSI and MOSEI 227
Table 6.3   Results on SIMS 228
Table 6.4   Dataset statistics for benchmark MSA dataset in format
             negative/neutral/positive 234
Table 6.5   AUILC results comparison with baseline models on MOSI
             and SIMS dataset 237
Table 6.6   TFR-Net results for different modality missing
             combinations 238
Table 7.1   Some of the supported features in M-SENA 244
Table 7.2   Statistics of the generalization ability test dataset, where "en"
             represents "English", "ch" represents "Chinese" 246
Table 7.3   Results for feature selection 247
Table 7.4   Experiment results for MSA benchmark comparison 248
Table 7.5   Results for English generalization ability test 251

Chapter 1
Overview

Abstract This chapter introduces the background and significance of sentiment analysis and multimodal machine learning. With the rapid development of online social media, various forms of information have covered the entire Internet, and people express their emotions in diverse ways. Therefore, the concept of multimodality emerges, providing different representations for each modality of information. At the same time, the rich evaluation information in multimodal forms enriches the evaluation of products, works, and other objects on various platforms, enabling businesses and users to access the desired information conveniently and accurately. This chapter also discusses the advancements in deep learning in fields such as natural language processing and computer vision, highlighting the importance of multitask learning in multimodal sentiment analysis. Multitask learning improves the robustness and learning performance of models by sharing model parameters and combining the characteristics of different tasks. This chapter aims to provide a systematic overview of recent research methods in multimodal machine learning, covering the perspectives of multimodal representation learning and multimodal fusion.

1.1 Overview of Multimodal Sentiment Analysis

Sentiment Analysis, also known as opinion mining, aims to analyze the sentiment tendencies or attitudes of features from data. It is widely used in many industries, including dialog generation, recommendation system, and so on [1]. With the rapid development of online social media, different modalities of information gradually cover the entire Internet, which is closely related to how people express their feelings, from the initial text comments such as plain text Emails, post bars, and forums to the later media rich in video and audio such as pictures and short videos. Therefore, the concept of multimodality came into being. For each modal of information, there are different physical carrier forms or representations, which are called modality. The rich evaluation information of users also enriched the evaluation information of commodities, works, and other objects on various platforms. Therefore, both businesses and users can get the information they want to know more conveniently and accurately according to the corresponding sentiment evaluation.

1.1.1   Overview of Research on Multimodal Sentiment Analysis

In the early stage, due to the limitation of information processing ability and the unitary modality of data, sentiment analysis mainly focuses on the analysis and processing of text, vision, and other single-modality data. With the sustaining development of information technology, the learning model represented by deep learning constantly refreshes the performance indicators of natural language processing, audio analysis, computer vision recognition, and many other fields. In this process, the sentiment analysis ability of single-modality content has also been significantly improved [2, 3]. In addition, with the continuous popularity of mobile Internet, various short video applications, such as TikTok and KUAISHOU, gradually come into being. Traditional text content can no longer meet people's daily needs, and more and more people are keen to use short video to record and share their life. Short video data is a typical multimodal data, including image sequence, audio, and words. Similarly, in practical application scenarios, the interaction between people and a service robot in the real scenario is also a typical multimodal information interaction. As shown in Fig. 1.1, the audio and text information in a video can be easily obtained after certain transformation and extraction. These changes make some researchers gradually realize that single-modality content has natural information limitations. In some cases, it is difficult to identify the true emotions of features only by relying on single-modality information. For example, the same sentence will convey different emotional tendencies under different intonation and facial movements. In this case, how to effectively integrate different modal data for comprehensive sentiment analysis has become an urgent problem to be solved [4]. Thus, Multimodal Sentiment Analysis (MSA) emerged. It should be emphasized that the multimodal information studied in this book consists of three types of single-modality information that are face picture sequence, audio, and text.

Different from single-modality sentiment analysis, MSA needs to process text, audio, and video data simultaneously. If the results of each single-modality sentiment analysis are simply combined, for example, the strategy of three votes is

a. Video Modality   b. Audio Modality   c. Text Modality

Fig. 1.1 Example of multimodal data

adopted, the complementarity between each modal data will not be fully explored, resulting in poor effects in practice. Therefore, the focus of MSA is how to fully explore the complementarity between different modalities and expand the effect gain of the fusion process. In addition, considering the temporal features of multimodal data, it is necessary to consider two dimensional relations in the modeling and analysis process, the relationship between different time periods in the same modality, and the relationship between different modalities in the same time period. Based on this, existing studies disassemble multimodal learning into five subproblems, which are Translation, Alignment, Representation, Fusion, and Colearning [5]. Among them, translation, alignment, and collaboration are highly related to the features of multimodal data and are not covered in this book.

The purpose of Representation Learning is to obtain highly complementary learning representations from each single modal data. This process is a crucial part of MSA, and the quality of representation will have an important impact on the fusion and classification effect in the later stage. Scholars represented by Bengio believed that a good representation should have multiple attributes at the same time, including smoothness, space-time coherence, sparsity, natural aggregation, and so on [6]. Among them, smoothness means that the more similar the representation is, the more similar the result should be, which is also a basic assumption of the current learning algorithm. Space-time coherence means that the representation of adjacent data in the time dimension should be adjacent in the space dimension, that is, the representation of the data in the later time should be the result of superimposed small amplitude deviation of the representation of the data in the previous time. Sparsity means that for a given observation data, the representation generated should be close to zero in most dimension Spaces and only play a role in a few feature dimensions, which is equivalent to the result of feature selection. Natural aggregation means that data representations of the same category should be clustered together in space. In addition, in multimodal learning, the consistency and difference between different modalities should be taken into account in addition to the above attributes, since the structure and content of data of different modalities are very different. For the same emotion category, different single-modality representations should retain a certain degree of similarity, and because the emotion category expressed by each single-modality data is not always consistent, different single-modality representations should also have a certain degree of difference.

The aim of Representation Fusion is to make full use of the complementarity between the different modal representations and to obtain a compact, informative representation. This process can best reflect the advantages of multimodal learning and is also a hot issue in related research. After the development in recent years, from simple horizontal and vertical splicing [7] to complex fusion mechanism based on tensor and attention mechanism [4, 8–10], researchers have designed various representational fusion structures and achieved good experimental results.

Considering the above two subproblems comprehensively, the model framework shown in Fig. 1.2 is a very typical multimodal sentiment analysis method [4, 8]. In this framework, because each single-modality representation learning subnetwork does not interfere with each other and the fusion process is in the latter part of the

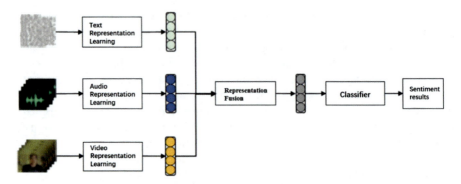

Fig. 1.2 A typical multimodal sentiment analysis framework

model, it is called as the late fusion framework of independent representation in this book. In order to make the multimodal fusion adequate and effective, it is necessary that the representation features learned by each single-modality subnetwork have sufficient complementarity. In a multimodal scenario, a modal depiction with rich information should contain two aspects of information: information on modal coherence and information on modal differences [11]. Among them, information on modal coherence refers to the common features of all modalities, emphasizing the commonality in data of different modalities. Information on modal differences refers to the unique features of each single modality and emphasizes the differences in data of different modalities. Since this book only considers the learning algorithm based on deep learning, it is necessary to combine the features of learning algorithms for method research and design in order to guide multimodal models to learn the single-modal representation that takes into account the two types of information. Generally, the learning algorithm includes two subprocesses: forward guidance and backward guidance. Forward guidance uses the framework structure of the model to convert the input data according to the preset way and get the final output result. Backward guidance promotes model parameters to advance toward predefined optimization aims through the backward propagation process of deep learning. From another perspective, the ultimate purpose of backward guidance is to correct the effectiveness of forward guidance. The existing research focuses on the design of various sophisticated representations and fusion structures, and pays more attention to the forward guidance of the model, but ignores the backward guidance. Considering the backward guidance process, the existing multimodal sentiment analysis models learn and train under the supervision of unified multimodal labeling. Therefore, unified multimodal labeling is not conducive to each single-modality substructure to learn differentiated information conforming to its own emotional properties.

Based on multimodal learning, people want to summarize information from all available modalities intuitively and build a learning model based on summarized information. Ideally, the model after learning should indicate the relative emphasis on different modalities for a particular task. This learning idea is the fusion of multiple modalities. Multimodal fusion is ubiquitous in existing multimodal technologies, including early and late fusion [12, 13], hybrid fusion [14], model integration [15], and, more recently, deep neural network-based joint training methods [16–18]. Each of these approaches combines elements (or intermediate elements) together and models them together to make decisions. Because of the modality of aggregation operation, such methods are called additive methods. But these fusion methods have recently hit a bottleneck. There are two main problems. Firstly, the multimodal data set itself has certain fuzziness. The emotion is artificially defined, and the data under the same label may not have high similarity, which makes it difficult for the machine to develop a model with high fitting degree according to the training. Secondly, there is a lot of variation between the different modal data. Taking text, audio, and video modalities as an example, how to integrate the features of the three modalities without losing information has become a big problem.

1.1.2 Overview of Related Research on Modality Loss

In addition, existing multimodal sentiment analysis models are often based on joint representations between learning modalities, and these methods have achieved impressive performance on multimodal sentiment benchmark data sets. However, real-time user data in real scenarios is often more complex. Firstly, the duration of each modality sequence varies with the sampling frequency of the modality sampler, leading to the nonalignment of modality features. As shown in Fig. 1.3, the speaker's expression of a word information often corresponds to a video frame and audio waveform, and different modal features often have different sequence lengths.

In addition, as shown in Fig. 1.4, many unavoidable factors, such as translation errors in user-generated videos, verbal representations of languages, words that are beyond the scope of the dictionary, background noise, sensor failures, and so on, may lead to the failure of the modal feature extractor, leading to random loss of modal features. How to resolve potential misalignment issues and missing features in

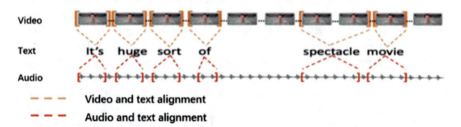
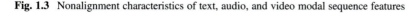

Fig. 1.3 Nonalignment characteristics of text, audio, and video modal sequence features

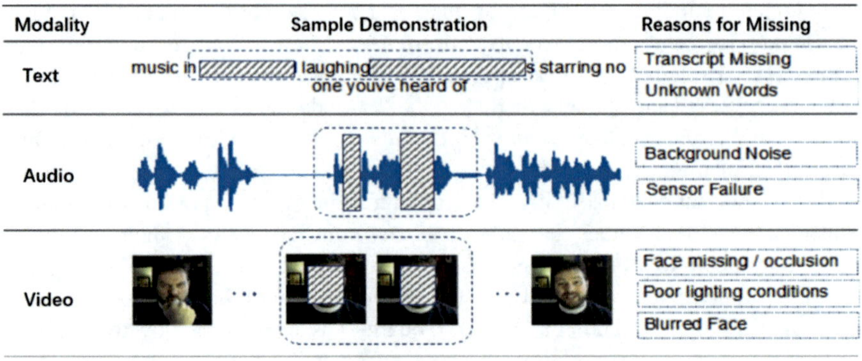

Modality	Sample Demonstration	Reasons for Missing
Text	music in ////////// laughing/////////////s starring no one youve heard of	Transcript Missing / Unknown Words
Audio		Background Noise / Sensor Failure
Video		Face missing / occlusion / Poor lighting conditions / Blurred Face

Fig. 1.4 Factors that may trigger modal random deficits

real-time multimodal data has become the core challenge of applying existing models to actual industry scenarios.

Based on this, this book will further discuss the problem of unaligned multimodal sentiment classification with random feature missing, design and implement a deep learning model based on feature reconfiguration, and improve the robustness of the model against random feature missing. At the same time, the book presents M-SENA, a multifunctional and multimodal platform for displaying sentiment analysis. This platform is used to complete model training, parameter back-up, model analysis, and end-to-end real-time video data assessment and testing. On the one hand, this book proposes a method for multimodal emotion classification with random feature missing, which improves the robustness of model representation learning to a certain extent. On the other hand, M-SENA can intuitively display the performance and classification basis of different models for real-time data, which is conducive to further model research and analysis.

1.1.3 Conclusion

Firstly, this section gives an overview of multimodal sentiment analysis, and gives a detailed introduction to the two key problems in the study of multimodal learning, multimodal representation learning, and multimodal representation fusion. Secondly, on the basis of the two subproblems above, the advantages and disadvantages of existing multimodal learning strategies are analyzed. In view of the shortcomings of existing multimodal learning strategies, this book will explore how to effectively use the previous multimodal coherence and the representation of differences for multimodal sentiment analysis tasks. In addition, this book systematically introduces the multimodal sentiment analysis based on deep learning, discusses and summarizes the application and the pros and cons of different algorithms, selects sentiment analysis cases, and demonstrates the effect of related algorithms, hoping to enlighten readers further.

1.2 Overview of Multimodal Machine Learning

In recent years, the field of machine learning has made great strides with the rapid development of deep learning. Since about 2010, the accuracy of large-scale Automatic Speech Recognition (ASR) using fully connected Deep Neural Networks (DNN) and Deep Automatic Encoders (DAE) has improved dramatically. In computer vision (CV), the deep convolutional neural network (CNN) model is used to make a series of breakthroughs in large-scale image classification tasks and large-scale target detection tasks. In Natural Language Processing (NLP), the semantic slot filling method based on Recursive Neural Networks (RNN) [19] has also reached a new level in spoken language understanding. In machine translation, sequential models based on the RNN encoder-decoder model and the attention mechanism model [20] also perform well.

Despite advances in visual, audio, and language processing, many research problems in AI involve more than one modality. For example, Intelligent Personal Assistants (IPAs) for understanding human communicative intentions, a task that involves not only spoken language, but also a combination of physical and pictorial language [21]. As a result, the study of multimodal modeling and learning methods has a broad range of meaning.

In this section, the intelligent models and learning methods of multimodal machine learning are reviewed. In order to provide a systematic overview, the current research methods of multimodal machine learning in recent years are elaborated from the perspective of multimodal representation learning and multimodal fusion.

1.2.1 Overview of Multimodal Representation Learning

Representation learning of multimodal data is the core problem of multimodal machine learning. The development of single-modal representations has been extensively studied in the field of machine learning. At present, most images are represented by learning by using neural structures (such as convolutional neural networks (CNN)) [22]. In the audio field, acoustic features such as Mel-Frequency Cepstral Coefficient (MFCC) have been replaced by data-driven deep neural networks for speech recognition [23] and recursive neural networks for language analysis [24]. In the process of natural language processing, the embedding of word context [25] is used to learn the representation of text features, which has replaced the representation of the number of word occurrences in the document, which was initially relied on. However, the representation of multiple data modes presents numerous challenges, such as how to combine data from different sources; how to manage different levels of noise; and what to do with lost data. Good representation is very important for the performance of machine learning models.

Based on extensive research work in the field of single-modal representation, the multimodal representation learning methods, which were based on correspondence between representations, can be divided into two types, Joint Representation and Coordinated Representation [5], in recent years.

Associative Representation Learning

Joint-type representations model three single-modal original representations jointly to obtain a single unified multimodal representation. In essence, the two phases of representation learning and representation fusion are combined into a single process. In this structure, the relationship between the individual modalities is not reciprocal and is usually divided into primary and secondary modalities according to their contribution to the result. Since textual content contains more information, textual modality is often regarded as the primary modality in multimodal tasks, while audio and video modalities are considered as auxiliary modalities. In 2019, Wang et al [26] proposed a regressive attention-based variational coding network, which essentially uses the audio and video representations to fine-tune the textual word vector representation, and to control the magnitude of the adjustment. We also introduced a linear gating unit to generate the adjusted weights. Subsequently, Rahman et al [27] borrowed this idea to a text pretraining language model and achieved better experimental results.

Collaborative Representation Learning

Collaborative representation learning treats each single-modal data equally, and each modal data is mapped to a different feature space according to its own characteristics. However, in modeling single-modal data, the influence of other modal data on the current modal learning process is often taken into account. In this section, the synergetic representation is further classified into strong synergetic representation and weak synergetic representation based on the presence or absence of this influence, and the latter can also be referred to as independent representation.

Since all single-modal information is temporal data, a large amount of research work on strongly collaborative representation learning then exploits the temporal consistency between different modal data to construct association models. In 2018, Zadeh et al [9] used three long short-term memory networks (LSTM) [28] to model the temporal dependencies between different modal data separately. Then, different modal information is superimposed within a word-level time window to obtain the multimodal fusion information in the current time period. Combined with temporal attention mechanism, the temporal dependency relationship of the multimodal fusion information is simulated. However, this type of model requires multimodal data that are aligned at the word-level, and has higher requirements for the content of the data. In 2019, Transformer [29], a temporal model based on the self-attentive mechanism, is more effective in Natural Language Processing because of its higher timeliness and better results. Tsai et al [30] firstly introduced this model to multimodal learning by setting modality A as the Query vector in Transformer and modality B as the Key and Value vectors, thus constructing a cross-modal attention mechanism from modality A to modality B. Numerous experiments show that this model achieves prediction results that far exceed previous models on both aligned and unaligned multimodal data.

Weak collaborative representation learning is a simple and widely used learning paradigm. Since the influence of other modal information is not considered in the single-modal learning process, this learning paradigm is essentially three basic single-modal representation learning models. The modular design advantage allows it to fully draw on research results from various single-modal domains. In text representation learning models, early researchers generally generate corresponding sentence representations based on word vectors from Word2Vec or GloVe, and then combine them with temporal models such as LSTM. In the last 2 years, large-scale pretrained language models represented by BERT [31] have gradually replaced the earlier structures and dominated in text representation learning.

In the representation learning model for audio and video, if the representation learning model in its specific domain is directly applied to the multimodal representation learning structure, two problems are raised. Firstly, the whole model structure is too heterogeneous, and the existing multimodal datasets are all small in size to adequately train such a complex network. Secondly, compared to text data, the prediction accuracy of audio and video data is easily disturbed by nonsubjective factors such as environment, which is not conducive to directly learning the representation features related to the actual task. Therefore, existing studies have used the structure of "pre-extracted features + re-learning" to build audio and video representation learning models. In practice, audio and video features that are highly relevant to a specific task are extracted from them by combining relatively mature feature extraction tools in the field of single-modal data (audio or video) analysis to form the original feature set. Further, the learned representations of audio and video are obtained using a combination of temporal and multilayer perceptron models.

Among the above multimodal representation learning classifications, weakly synergistic representations are the main class of representation learning methods studied in this book. Therefore, after obtaining each single-modal feature representation, it is immediately followed by representation fusion, that is, fusing multiple single-modal representations into one multimodal representation. The current state of the method research will be described in detail in the next section.

1.2.2   Overview of Multimodal Representation Fusion

Multimodal fusion is one of the most studied aspects of multimodal machine learning and one of the most fundamental subproblems in multimodal learning, which aims to fully exploit the complementarity between multimodal data to further improve the robustness of prediction results. In 2003, in multimodal tasks, researchers have verified that the representation of multiple modal fusions can effectively improve the robustness of the model [32]. Multimodal fusion can be classified as prefusion, mid-fusion, postfusion, and end-fusion according to the stage in which the multimodal fusion is located in the model.

Prefusion combines the original features of each single-modal modality at the data input stage. Mid-fusion implements the cross-modal fusion process during the learning of the single-modal representation. Postfusion follows immediately after the learning of each single-modal representation. End-fusion is the aggregation of the results after the results of each single-modal analysis have been obtained. In addition, it is also possible to combine multiple stages of fusion in the model at the same time, and this approach is called hybrid fusion.

Prefusion

Prefusion refers to the fusion of multiple independent datasets into a single feature vector, which is fed into a machine learning classifier. Since the front-end fusion of multimodal data often does not fully utilize the complementarity among multiple modal data, and the original data fused at the front-end usually contains a large amount of redundant information. Therefore, multimodal front-end fusion methods are often combined with feature extraction methods to remove redundant information, such as principal component analysis (PCA), maximum correlation minimum redundancy algorithm (mRMR), autoencoders, and so on.

Mid-Fusion

Mid-fusion refers to converting different modal data into high-dimensional feature representations first, and then fusing them with the intermediate layer of the model. Taking neural networks as an example, *mid-fusion* first uses neural networks to transform the original data into a high-dimensional feature representation, and then obtains the commonality of different modal data in the high-dimensional space. A major advantage of the intermediate fusion method is the flexibility to choose the location of fusion.

Postfusion

Postfusion refers to the fusion of each single-modal representation immediately after it is learned, and feature stitching is a simple and effective postfusion method that directly expands the feature size for subsequent tasks by stitching multiple single-modal features together. However, this approach treats individual modal features independently and ignores the dynamic interactivity features between modalities. In 2017, Zadeh et al [4] proposed a Tensor Fusion Network (TFN) to capture the correlation features between modalities by performing multidimensional fork product operations on features. However, this high-order tensor operation causes the size of the fused features to increase geometrically, resulting in many redundant features and increasing the computational overhead of subsequent tasks.

In 2018, Liu et al [8] proposed a low-rank tensor fusion (LMF) model based on TFN: this method introduces three modal specificity factors and performs the fusion operation in the second-order space, which significantly reduces the number of fused representation features and effectively solves the feature redundancy and operation overhead in TFN, the problem of large operational overhead. Further, in 2020, Sahay et al [33] extended this fusion method to Transformer-based model architectures, which also achieved good performance. However, such tensor operation-based fusion methods are poorly interpreted and do not yield the contribution of each single-modal representation to the fused features. Thus, Zadeh et al [10] proposed a dynamic fusion graph structure model. This structure uses a hierarchical fusion approach with a three-way fusion on top of a two-by-two fusion, and introduces learning-based weights for referring to the contribution of different single-modal features to the fused features. Subsequently, some researchers have also designed more complex multimodal fusion structures based on the above work.

End-Fusion

End-fusion refers to fusion using fusion mechanisms such as averaging, voting schemes, weighting based on channel noise and signal variance, or learning models. It allows the use of different models for each mode, as different predictors can better model each individual mode, thus allowing greater flexibility. In addition, it makes prediction easier when one or more modalities are missing, even allowing training when no parallel data is available.

The multimodal machine learning methods proposed subsequently in this book mainly consider weakly collaborative type representations, so this subsection focuses on the state of the art of methods coming to late fusion.

1.2.3 Conclusion

This section presents a review of the research related to multimodal representation and multimodal fusion in multimodal machine learning, respectively, and analyzes the advantages and disadvantages of each approach in detail. Nowadays, the mainstream learning-based algorithms consist of two processes: forward guidance, which constrains the input data toward a predefined goal through a predefined model structure, and backward guidance, which updates the parameters of the model driven by an optimization goal. In the above research work, it is expected that the individual single-modal representation learning structures are used to obtain modal discrepancy information and modal fusion structures to obtain modal consistency information. It is easy to see that all these works only consider the process of forward guidance and ignore the design of backward guidance. The core of backward bootstrapping lies in the design of optimization objectives, and the aforementioned work contains only multimodal-level loss optimization, but multimodal-level supervision values are not always applicable to the learning process of single-modal representation, thus making it easy to guide the representation learning model to learn more modal consistency information to the detriment of learning modal discrepancy information. To address the above shortcomings of multimodal representation and multimodal fusion methods, this book explores how to effectively use multitask learning mechanisms to add additional single-modal-level optimization goals in addition to multimodal-level optimization goals to guide the model to learn modal discrepancy information. The model is guided by multiple optimization objectives, a typical multitask learning paradigm.

1.3 Overview of Multitask Learning Mechanisms

Multitask Learning (MTL) is a branch of machine learning that learns multiple tasks simultaneously through a shared model. This structure can combine the characteristics of different tasks and improve the robustness of the model and the learning effect by sharing the model parameters [34]. Multitask learning methods have the advantages of improving data efficiency, reducing overfitting through shared representation, and learning quickly using auxiliary information. In multimodal sentiment analysis, multitask learning can be used to better integrate the three modalities. In general, the underlying network parameters of multiple tasks are related to each other, while the top layers are independent of each other. During the training process, all or some of the parameters of the underlying layer are jointly optimized by multiple tasks, thus achieving the goal of joint training of multiple tasks. Depending on the difference in how the underlying parameters are shared, they can be classified as Hard Sharing and Soft Sharing. In the former, the underlying parameters are fully shared, while in the latter, they are partially shared.

1.3.1   Multitasking Architecture in Computer Vision

In single-task settings, many significant developments in computer vision architectures have focused on new network components and connections to improve optimization and extract more meaningful features, such as batch normalization [35], residual networks [36], and squeeze and excitation blocks [37]. In contrast, many multitasking architectures for computer vision focus on dividing the network into task-specific shared components in a way that allows generalization through shared and task-to-task information flows while minimizing negative transfer.

The basic feature extractor consists of a series of convolutional layers that are shared among all tasks, and the extracted features are used as input to the task-specific output header.

A lot of work [38–42] proposes architectures that are variants of the shared backbone. Zhang et al [39] were the first team to carry out related work in these works, and their work introduces task-constrained deep convolutional networks (TCDCN), whose architecture is shown in Fig. 1.5. We propose to improve the performance of face landmark detection tasks by jointly learning head pose estimation and facial attribute inference. The architecture of multitask network cascaded MNCs [40] is similar to that of TCDCNs, but with one important difference: the output of each task-specific branch is appended to the input of the next task-specific branch, forming a cascading flow of information.

Not all MTL architectures for computer vision in related studies include a shared global feature extractor with task-specific output branches or modules. Some work has taken a separate approach [43–45]. Instead of a single shared extractor, these architectures provide separate networks for each task with information flow between parallel layers in the task network. The Cross-Stitch network architecture [43] in Fig. 1.6 depicts this idea. The Cross-Stitch network consists of a single network for each task, but the input to each layer is a linear combination of the outputs of the previous layer of each task network. The weights of each linear combination are learned and task-specific so that each layer can choose which tasks to utilize information from.

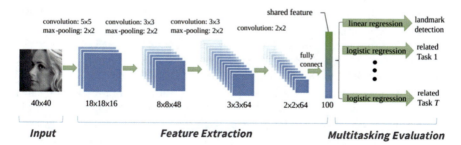

Fig. 1.5 The architecture of TCDCN

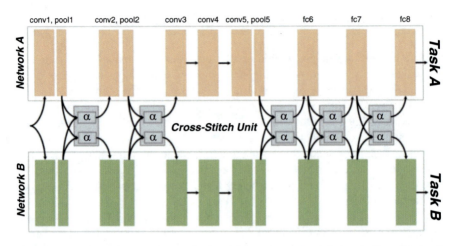

Fig. 1.6 Cross-Stitch network structure. Each task has a separate network, but the Cross-Stitch unit linearly combines information from parallel layers of different task networks

1.3.2   Multitasking Architecture in Natural Language Processing

Natural language processing is naturally applicable to MTL, because one can ask a large number of relevant questions about a given piece of text, as well as the task-independent representations often used in modern NLP techniques. In recent years, the development of NLP neural architectures has gone through several phases, with traditional feed-forward architectures evolving into recurrent models and recurrent models being replaced by attention-based architectures. These phases are reflected in the application of these NLP architectures to MTL. It should also be noted that many NLP techniques can be considered multitasking, because they construct task-independent general representations (e.g., word embeddings), and that a discussion of multitasking NLP under this interpretation would include a large number of more widely known approaches that are general-purpose NLP techniques. Here, for the sake of practicality, we restrict the discussion to techniques that primarily include explicit learning of multiple tasks simultaneously, with the ultimate goal of performing these tasks simultaneously.

Early work has used traditional feed-forward (nonattention-based) architectures for multitask NLP [46–48]. Many of these architectures have structural similarities to earlier shared architectures for computer vision: a shared global feature extractor followed by task-specific output branches. However, in this case, the features are word representations. Collobert and Weston [48] use a shared lookup table layer to learn word representations, where the parameters of each word vector are learned directly by gradient descent. The rest of the architecture is task-specific and consists of convolution, time-varying max, fully connected layers, and Softmax outputs.

The introduction of modern recurrent neural networks has produced a new family of models for multitask NLP, introducing new recursive architectures [49–51]. Sequence-to-sequence learning [20] was applied to the multitask learning of Luong et al [50] In this work, we explored three variants of parameter sharing schemes for multitask seq2seq models, which they named one-to-many, many-to-one, and many-to-many. In the one-to-many model, the encoder is shared by all tasks and the decoder is task-specific. This is useful for handling tasks that require different formats of output, such as translating a text into multiple target languages. In many-to-one, the encoder is task-specific and the decoder is shared. This is in contrast to the usual parameter sharing scheme where the earlier layers are shared and made available to task-specific branches. The many-to-one variant is applicable when the set of tasks needs to be output in the same format, such as image captioning and machine translation into the same target language. Finally, we explore many-to-many variants where there are multiple shared or task-specific encoders and decoders.

So far, the subarchitectures corresponding to each task are symmetric in all the NLP architectures discussed above. In particular, the output branch of each task appears at the maximum network depth of each task, which means that the supervision of each task-specific feature occurs at the same depth. Related work [52–54] suggests supervising "low-level" tasks in the early layers so that features learned for these tasks can be used for higher-level tasks. By doing so, a clear task hierarchy can be formed and provides a direct way for information from one task to help solve another task. This can be used for iterative reasoning and feature combination in a template called cascade information, illustrated in Fig. 1.7.

Despite the popularity of Transformers' bi-directional encoder representation (BERT) [31], there are few applications of the MTL text encoding approach. Liu et al [55] extended the work of the literature [47] by adding a shared BERT embedding layer to the architecture. The overall network architecture is very similar to the literature [47], with the only difference being the addition of a BERT contextual embedding layer after the input embedding vector in the original architecture. This new MTL architecture called MT-DNN achieves SOTA performance in the GLUE task [56], and the related results were published with eight out of nine tasks completed.

Each modality has a separate network to process the input and the aggregated output is processed by an encoder-decoder called the central neural processor. The output of the CNP is then passed to several task-specific output headers.

1.3.3   Multitasking Architecture in Multimodal Learning

Multimodal learning is an interesting extension of many of the motivating principles behind multitask learning: sharing representations across domains reduces overfitting and improves data efficiency. In the multitask single-modal case, representations are shared across multiple tasks, but within a single modality. However, in

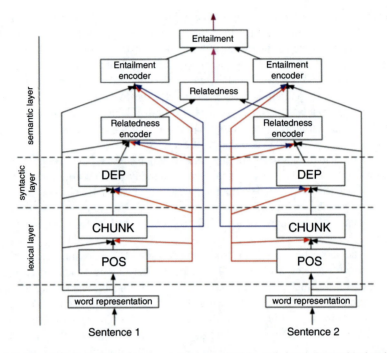

Fig. 1.7 Cascade information of various tasks supervised in each layer. Lower-level tasks are supervised in the earlier layers

the multitask multimodal case, the representation is shared across tasks and modalities, which provides another abstraction layer through which the learned representation must be generalized. This suggests that multitask multimodal learning may enhance the advantages already presented by multitask learning.

Nguyen and Okatani [57] introduced an architecture for sharing visual and verbal tasks by using dense coattentive layers [58], where tasks are organized into a hierarchy and lower-level tasks are supervised in earlier layers. This develops dense common attention layers for visual quizzing, especially for integrating visual and linguistic information. The approach is to search the layers for each task to understand the task hierarchy. The architecture of Akhtar et al [59] processes visual, audio, and textual inputs to classify sentiment and emotion in human speaker videos, using a bidirectional GRU layer and a pairwise attention mechanism for each pair of patterns to learn a shared representation containing all input patterns.

These works [57, 59] focus on a set of tasks, all of which share the same fixed set of patterns. In contrast, other works [60, 61] focus on building a generic multimodal multitasking model where one model can handle multiple tasks with different input domains. The introduced architecture [61] consists of an input encoder, an I/O mixer, and an autoregressive decoder. Each of these three blocks consists of a convolution, an attention layer, and a sparse expert mixture layer. We also demonstrated that extensive sharing between tasks can significantly improve the performance of

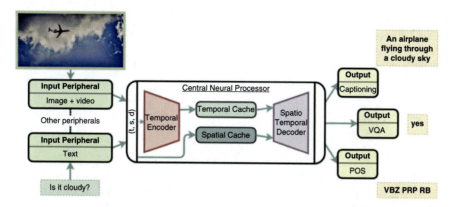

Fig. 1.8 The proposed OmniNet architecture [57]

learning tasks with limited training data. Instead of using the aggregation mechanisms found in various deep learning models, Pramanik et al [60] introduced an architecture called OmniNet, which has a spatiotemporal caching mechanism to learn dependencies across the spatial and temporal dimensions of the data. Figure 1.8 shows a relational graph. Each input pattern has a corresponding peripheral network, and the outputs of these networks are aggregated and fed to the central neural processor, whose outputs are fed to the task-specific output head. The CNP has an encoder-decoder architecture with spatial caching and temporal caching. OmniNet achieves performance comparable to SOTA methods for POS tagging, image captioning, visual question answering, and video activity recognition. Recently, Lu et al [62] introduced a multitasking model that can process 12 different datasets simultaneously, aptly named 12-in-1. Their model achieves better performance than the corresponding single-tasking model on 11 of the 12 tasks and achieves the best performance on 7 tasks using multitasking training as a pretraining step. The architecture is based on the ViLBERT model [62] and is trained using a mixture of dynamic task scheduling, course learning, and hyperparametric heuristics.

Multitask learning is a machine learning approach that has started to regain popularity in recent years, which shares some or all of the parameters of multiple related single-modal tasks for training, increasing the generalization ability of the trained model for each single-modal task while expanding the size of the dataset and alleviating the data sparsity problem. Specifically, the multitask learning subtask in this section is the learning of sentiment classification for each single-modal branch. Unlike traditional multitask learning, the training data for multitask-based multimodal sentiment classification is identical across subtasks, with only the single-modal labels varying. And because of these differences, the two bottlenecks mentioned in the previous section are ideally solved: firstly, for the problem of inconsistent modal representation of the dataset, the discrepancy is not solved by adding the labels of individual modalities, but from another perspective, the original single-task data is subdivided twice, which makes the discrepancy between similar data after subdivision narrower and facilitates better supervised learning by the machine. Second, for the problem of feature fusion difficulties, the emergence of subtask branching allows the single-modal features to better maintain their own characteristics during each iteration of learning, thus expanding the variability between modal features. In this way, the fused features can play their proper role.

1.3.4 Conclusion

This section has explored an important branch of machine learning—multitasking learning mechanisms. The main two areas of computer vision and natural language processing are introduced in a comprehensive manner, and finally, the multitasking architecture in multimodal learning is introduced in detail, and the advantages and disadvantages of existing research methods are analyzed. A variety of methods have been proposed by previous authors, but the results are not satisfactory. This is because there are bottlenecks in the existing multimodal sentiment classification tasks that need to be solved. For the bottlenecks in multimodal learning at present, this book proposes a method that combines multitasking mechanisms to multimodal sentiment classification.

1.4 Summary

This chapter provides an overview of multimodal sentiment analysis, focusing on two key problems: multimodal representation learning and multimodal representation fusion. The existing strategies and their limitations are analyzed, leading to an exploration of effectively utilizing multimodal coherence and representation differences for sentiment analysis tasks. This chapter further delves into deep learning-based multimodal sentiment analysis, discussing various algorithms, their applications, and their pros and cons. Sentiment analysis cases are presented to demonstrate the effectiveness of the algorithms, aiming to enlighten readers. Additionally, this chapter reviews research related to multimodal representation and fusion, highlighting the importance of both forward and backward guidance in learning algorithms. It points out the limitations of existing approaches, particularly the lack of backward guidance design. To address these shortcomings, this chapter proposes adding single-modal-level optimization goals to guide the learning of modal discrepancy information. This is achieved through multitask learning mechanisms, introducing multiple optimization objectives. Furthermore, this chapter explores multitasking learning mechanisms as an important branch of machine learning. It comprehensively introduces the main areas of computer vision and natural language processing and analyzes the advantages and disadvantages of existing research methods. Considering the current bottlenecks in multimodal sentiment classification tasks, this chapter proposes a multitasking mechanism that combines different tasks to enhance multimodal sentiment classification. In summary, this chapter provides a comprehensive overview of multimodal sentiment analysis, explores the limitations of existing strategies, proposes new approaches, and highlights the importance of multitasking learning mechanisms for effective multimodal sentiment classification.

References

1. Bing L (2012) Sentiment analysis and opinion mining. Synth Lect Hum Lang Technol 5(1): 1–167
2. Yadollahi A, Shahraki AG, Zaïane OR (2017) Current state of text sentiment analysis from opinion to emotion mining. Assoc Comput Machin Comput Surv 50(2):1–33
3. Shan L, Weihong D (2021) Deep facial expression recognition: a survey. IEEE Trans Affect Comput 13(3):1195–1215
4. Zadeh A, Chen M, Poria S, et al (2017) tensor fusion network for multimodal sentiment analysis. Proceedings of the 2017 Association for Computational Linguistics Conference on Empirical Methods in Natural Language Processing, 1103–1114
5. Baltrusaitis T, Ahuja C, Morency L-P (2019) Multimodal machine learning: a survey and taxonomy. IEEE Trans Pattern Anal Mach Intell 41(2):423–443
6. Bengio Y, Courville A, Vincent P (2013) Representation learning: a review and new perspectives. IEEE Trans Pattern Anal Mach Intell 35(8):1798–1828
7. Williams J, Comanescu R, Radu O, et al (2018) DNN multimodal fusion techniques for predicting video sentiment. Proceedings of Grand Challenge and Workshop on Human Multimodal Language, 64–72
8. Liu Z, Shen Y, Lakshminarasimhan VB, et al (2018) Efficient low-rank multimodal fusion with modality-specific factors. Proceedings of the 56th Annual Meeting of the Association for Computational Linguistics, (1):2247–2256
9. Zadeh A, Liang PP, Mazumder N, et al (2018) Memory fusion network for multi-view sequential learning. Proceedings of the 32nd Association for the Advancement of Artificial Intelligence Conference on Artificial Intelligence, 5634–5641
10. Zadeh AB, Liang PP, Poria S, et al (2018) Multimodal language analysis in the wild: Cmu-mosei dataset and interpretable dynamic fusion graph. Proceedings of the 56th Annual Meeting of the Association for Computational Linguistics, (1): 2236–2246
11. Hazarika D, Zimmermann R, Poria S (2020) Misa: Modality-invariant and-specific representations for multimodal sentiment analysis. Proceedings of the 28th Association for Computing Machinery International Conference on Multimedia, 1122–1131
12. Gunes H, Piccardi M (2005) Affect recognition from face and body: Early fusion vs. late fusion. Proceedings of the 2005 Institute of Electrical and Electronics Engineers international conference on systems, man and cybernetics, (4): 3437–3443
13. Snoek CG, Worring M, Smeulders AW (2005) Early versus late fusion in semantic video analysis. Proceedings of the 13th annual Association for Computing Machinery international conference on Multimedia, 399–402
14. Atrey PK, Anwar Hossain M, El Saddik A et al (2010) Multimodal fusion for multimedia analysis: a survey. Multimedia Systems 16(6):345–379
15. Dietterich TG (2000) Ensemble methods in machine learning. Proceedings of the 1st International Workshop on Multiple Classifier Systems, 1–15
16. Wöllmer M, Metallinou A, Eyben F, et al (2010) Context-sensitive multimodal emotion recognition from speech and facial expression using bidirectional lstm modeling. Proceedings of InterSpeech, 2362–2365
17. Neverova N, Wolf C, Taylor G et al (2015) Moddrop: adaptive multi-modal gesture recognition. IEEE Trans Pattern Anal Mach Intell 38(8):1692–1706

18. Ngiam J, Khosla A, Kim M, et al (2011) Multimodal deep learning. Proceedings of the 28th International Conference on International Conference on Machine Learning, 689–696
19. Grégoire M, Dauphin Y, Yao K et al (2015) Using recurrent neural networks for slot filling in spoken language understanding. IEEE/ACM Trans Audio Speech Lang Process 23(3):530–539
20. Sutskever I, Vinyals O, Le QV, et al (2014) Sequence to sequence learning with neural networks. Proceedings of the 27th International Conference on Neural Information Processing Systems, 3104–3112
21. Shum H-Y, He X-D, Li D (2018) From Eliza to XiaoIce: challenges and opportunities with social chatbots. Front Informat Technol Electr Eng 19(1):10–26
22. Krizhevsky A, Sutskever I, Hinton GE (2012) Imagenet classification with deep convolutional neural networks. Commun ACM 60:84–90
23. Hinton G, Deng L, Dong Y et al (2012) Deep neural networks for acoustic modeling in speech recognition: the shared views of four research groups. IEEE Signal Process Mag 29(6):82–97
24. George T, Fabien R, Raymond B, et al (2016) Adieu features? end-to-end speech emotion recognition using a deep convolutional recurrent network. Proceedings of the 2016 Institute of Electrical and Electronics Engineers international conference on acoustics, speech and signal processing, 5200–5204
25. Mikolov T, Sutskever I, Chen K, et al (2013) Distributed representations of words and phrases and their compositionality. Proceedings of the 26th International Conference on Neural Information Processing Systems, 3111–3119
26. Wang Y, Shen Y, Liu Z et al (2019) Words can shift: dynamically adjusting word representations using nonverbal behaviors. Proc Assoc Advanc Artific Intellig Conf Artific Intellig 33(1): 7216–7223
27. Rahman W, Hasan M K, Lee S, et al (2020) Integrating multimodal information in large pretrained transformers. Proceedings of the 58th Annual Meeting of the Association for Computational Linguistics, 2359–2369
28. Hochreiter S, Schmidhuber J (1997) Long short-term memory. Neural Comput 9(8):1735–1780
29. Vaswani A, Shazeer N, Parmar N, et al (2017) Attention is all you need. Proceedings of the 31st International Conference on Neural Information Processing Systems, 5998–6008
30. Tsai YHH, Bai S, Liang PP, et al (2019) Multimodal transformer for unaligned multimodal language sequences. Proceedings of the 57th Annual Meeting of the Association for Computational Linguistics, 6558–6569
31. Devlin J, Chang MW, Lee K, et al (2019) Bert: Pre-training of deep bidirectional transformers for language understanding. Proceedings of the 2019 Conference of the North American Chapter of the Association for Computational Linguistics: Human Language Technologies, (1):4171–4186
32. Potamianos G, Neti C, Gravier G et al (2003) Recent advances in the automatic recognition of audiovisual speech. Proc IEEE 91(9):1306–1326
33. Sahay S, Okur E, Kumar SH, et al (2020) Low rank fusion based transformers for multimodal sequences. Proceedings of the 2nd Grand-Challenge and Workshop on Multimodal Language, 29–34
34. Zhang Y, Yang Q (2022) A survey on multi-task learning. IEEE Trans Knowl Data Eng 34(12): 5586–5609
35. Ioffe S, Szegedy C (2015) Batch normalization: Accelerating deep network training by reducing internal covariate shift. Proceedings of the 32nd International Conference on International Conference on Machine Learning, 37: 448–456
36. He K, Zhang X, Ren S, et al (2016) Deep residual learning for image recognition. Proceedings of the 2016 Institute of Electrical and Electronics Engineers conference on computer vision and pattern recognition, 770–778
37. Jie H, Shen L, Albanie S et al (2019) Squeeze-and-excitation networks. IEEE Trans Pattern Anal Mach Intell 42(8):2011–2023

38. Liu S, Johns E, Davison AJ (2019) End-to-end multi-task learning with attention. Proceedings of the Institute of Electrical and Electronics Engineers/Computer Vision Foundation Conference on Computer Vision and Pattern Recognition, 1871–1880
39. Zhang Z, Luo P, Loy CC, et al (2014) Facial landmark detection by deep multi-task learning. Proceedings of the 2014 European conference on computer vision, 94–108
40. Dai J, He K, Sun J (2016) Instance-aware semantic segmentation via multi-task network cascades. Proceedings of the Institute of Electrical and Electronics Engineers conference on computer vision and pattern recognition, 3150–3158
41. Ma J, Zhao Z, Yi X, et al (2018) Modeling task relationships in multi-task learning with multi-gate mixture-of-experts. Proceedings of the 24th Association for Computing Machinery Special Interest Group on Knowledge Discovery and Data Mining, 1930–1939
42. Zhao X, Li H, Shen X, et al (2018) A modulation module for multi-task learning with applications in image retrieval. Proceedings of the European Conference on Computer Vision, 401–416
43. Misra I; Shrivastava A; Gupta A, et al (2016) Cross-stitch networks for multi-task learning. Proceedings of the Institute of Electrical and Electronics Engineers conference on computer vision and pattern recognition, 3994–4003
44. Ruder S, Bingel J, Augenstein I, et al (2019) Latent multi-task architecture learning. Proceedings of the 33rd Association for the Advancement of Artificial Intelligence Conference on Artificial Intelligence and 31st Innovative Applications of Artificial Intelligence Conference and 9th Association for the Advancement of Artificial Intelligence Symposium on Educational Advances in Artificial Intelligence, 4822–4829
45. Yuan Gao, Jiayi Ma, Mingbo Zhao, et al (2019) DDR-CNN: Layerwise feature fusing in multi-task cnns by neural discriminative dimensionality reduction. Proceedings of the Institute of Electrical and Electronics Engineers/Computer Vision Foundation Conference on Computer Vision and Pattern Recognition, 3205–3214
46. Collobert R, Weston J, Bottou L et al (2011) Natural language processing (almost) from scratch. J Mach Learn Res 12:2493–2537
47. Liu X, Gao J, He X, et al (2015) Representation Learning Using Multi-Task Deep Neural Networks for Semantic Classification and Information Retrieval. Proceedings of the 2015 Conference of the North American Chapter of the Association for Computational Linguistics: Human Language Technologies, 912–921
48. Collobert R, Weston J (2008) A unified architecture for natural language processing: Deep neural networks with multitask learning. Proceedings of the 25th international conference on Machine learning, 160–167
49. Dong D, Wu H, He W, et al (2015) Multi-task learning for multiple language translation. Proceedings of the 53rd Annual Meeting of the Association for Computational Linguistics and the 7th International Joint Conference on Natural Language Processing, (1): 1723–1732
50. Minh Thang Luong, Quoc V Le, Ilya Sutskever, et al Multi-task sequence to sequence learning. Proceedings of the 4th International Conference on Learning Representations, 2016
51. Liu P, Qiu X, Huang X (2016) Recurrent neural network for text classification with multi-task learning. Proceedings of the 25th International Joint Conference on Artificial Intelligence, 2873–2879
52. Søgaard A, Goldberg Y (2016) Deep multi-task learning with low level tasks supervised at lower layers. Proceedings of the 54th Annual Meeting of the Association for Computational Linguistics, (2): 231–235
53. Sanh V, Wolf T, Ruder S. (2019) A hierarchical multi-task approach for learning embeddings from semantic tasks. Proceedings of the Association for the Advancement of Artificial Intelligence Conference on Artificial Intelligence, 853:6949–6956
54. Hashimoto K, Xiong C, Tsuruoka Y, et al (2017) A joint many-task model: Growing a neural network for multiple nlp tasks. Proceedings of the 2017 Conference on Empirical Methods in Natural Language Processing, 1923–1933

55. Liu X, He P, Chen W, et al (2019) Multi-task deep neural networks for natural language understanding. Proceedings of the 57th Annual Meeting of the Association for Computational Linguistics. 4487–4496
56. Wang A, Singh A, Michael J, et al (2018) GLUE: A multi-task benchmark and analysis platform for natural language understanding. Proceedings of the 2018 Empirical Methods in Natural Language Processing Workshop BlackboxNLP: Analyzing and Interpreting Neural Networks for NLP. 353–355
57. Nguyen DK, Okatani T (2019) Multi-task learning of hierarchical vision-language representation. Proceedings of the Institute of Electrical and Electronics Engineers/Computer Vision Foundation Conference on Computer Vision and Pattern Recognition, 10492–10501
58. Nguyen DK, Okatani T (2018) Improved fusion of visual and language representations by dense symmetric co-attention for visual question answering. Proceedings of the Institute of Electrical and Electronics Engineers/Computer Vision Foundation Conference on Computer Vision and Pattern Recognition, 6087–6096
59. Akhtar MS, Chauhan D, Ghosal D, et al (2019) Multi-task learning for multi-modal emotion recognition and sentiment analysis. Proceedings of the 2019 Conference of the North American Chapter of the Association for Computational Linguistics: Human Language Technologies, (1): 370–379
60. Subhojeet Pramanik, Priyanka Agrawal, Aman Hussain. Omninet: A unified architecture for multi-modal multi-task learning. arXiv preprint arXiv:1907.07804, 2019
61. Lukasz Kaiser, Aidan N. Gomez, Noam Shazeer, et al One model to learn them all.. arXiv preprint arXiv:1706.05137, 2017
62. Jiasen L, Goswami V, Rohrbach M, et al (2020) 12-in-1: multi-task vision and language representation learning. Proceedings of the Institute of Electrical and Electronics Engineers/ Computer Vision Foundation Conference on Computer Vision and Pattern Recognition, 10437–10446

Chapter 2
Multimodal Sentiment Analysis Data Sets and Preprocessing

Abstract This chapter emphasizes the importance of standard datasets in multimodal sentiment analysis. Deep learning algorithms have significantly improved the performance of multimodal sentiment analysis, making it a popular method for data analysis and prediction across various domains. High-quality manually annotated data helps researchers focus on algorithm development. Recent advancements in deep learning-based multimodal sentiment analysis are attributed to the contribution of datasets. This chapter provides an overview of publicly available multimodal sentiment analysis datasets, with a focus on those containing text, audio, and video modalities. It introduces CH-SIMS, a new Chinese dataset that includes multimodal and single-modal annotations for detailed and comprehensive analysis. This chapter also presents a multimodal multitask learning framework based on SIMS, which enhances performance by sharing feature representations and employing fusion models. The limitations of existing corpora are discussed, leading to the construction of CH-SIMS v2.0, a high-quality dataset with supervised and unsupervised data, simulating real-world human-computer interaction scenarios. This dataset enables improved acoustic and visual representation learning, as well as fine-grained emotion intensity prediction, enhancing the accuracy and effectiveness of multimodal sentiment analysis.

2.1 Multimodal Sentiment Analysis Datasets

2.1.1 Introduction

In 2006, Geoffrey Hinton introduced the concept of deep learning [1], which has continuously improved performance in various fields and has become the most popular method for data analysis and prediction. A large number of standard datasets are the foundation of its success. In supervised learning, manually annotated high-quality data helps researchers focus on algorithm development. In recent years, multimodal sentiment analysis algorithms based on deep learning have rapidly developed, thanks to the contribution of datasets. Therefore, this section briefly introduces standard datasets in this field. Table 2.1 summarizes the basic information

Table 2.1 Summary of public multimodal emotional analysis dataset

Dataset	Size	Language	Unimodal label	Multimodal label	Year
CMU-MOSI	2199	English	Not	Yes	2016
CMU-MOSEI	23,453	English	Not	Yes	2018
IEMOCAP	10,000	English	Not	Yes	2008
MELD	13,708	English	Not	Yes	2019

Table 2.2 Baseline dataset settings

Dataset	Training set	Valid set	Testing set	Summary
CMU-MOSI	1284	229	686	2199
CMU-MOSEI	16,326	1871	4659	22,856
IEMOCAP	3441	849	1241	5531
MELD	9989	1109	2610	13,708

of publicly available multimodal sentiment analysis datasets. It should be emphasized that this section only considers multimodal sentiment analysis datasets that include text, audio, and video modal information. Table 2.2 provides the training, validation, and test set partitioning for the reference datasets.

2.1.2 CMU-MOSI

The CMU-MOSI [2] dataset is a corpus for studying the sentiment and subjectivity of opinion videos in online sharing sites, such as YouTube. The dataset proposes a subjectivity annotation scheme to identify and segment different viewpoints expressed by speakers, enabling fine-grained viewpoint segmentation of online multimedia content. Using this scheme, 3702 video segments, including 2199 opinion segments, were reliably identified in the MOSI dataset. A total of 93 videos were randomly selected, and the final video set includes 89 different speakers, including 41 females and 48 males, most of whom are between the ages of 20 and 30. Although these speakers come from different racial backgrounds, such as White, African American, Hispanic, and Asian, they all express themselves in English, and these videos are from the United States or the United Kingdom. The emotional intensity is defined on a linear scale from -3 to $+3$, ranging from strongly negative to strongly positive. These intensity annotations were performed by online workers on the Amazon Mechanical Turk website.

2.1.3   CMU-MOSEI

CMU-MOSEI [3] is a large dataset used for researching multimodal emotion and sentiment recognition. It includes 23,453 annotated video segments from 1000 different speakers and 250 topics. Each video segment contains manually transcribed text aligned to phoneme-level audio. This dataset is very useful for expanding the scope of human multimodal language research in the field of natural language processing.

2.1.4   IEMOCAP

IEMOCAP [4] is a dataset consisting of acted, multimodal, and multispeaker data, collected during dyadic sessions where actors perform improvisations or scripted scenarios. The dataset includes video, speech, motion capture of face, and text transcriptions. Multiple annotators have labeled the dataset with categorical labels such as anger, happiness, sadness, neutrality, as well as dimensional labels such as valence, activation, and dominance. The dataset is large and provides detailed motion capture information and interactive settings that can elicit authentic emotional expressions. Therefore, IEMOCAP is a valuable addition to the existing databases in the community for the study and modeling of multimodal and expressive human communication.

2.1.5   MELD

Multimodal Emotion Lines Dataset (MELD) [5] is a multimodal emotional dataset that is enhanced and extended based on the Emotion Lines [6] dataset. MELD shares the same dialogue instances with Emotion Lines, but it adds information from the audio and visual modalities while retaining the textual information. MELD contains over 1400 dialogues and 13,000 utterances from the TV show Friends, in which multiple speakers are involved in the conversation. Each utterance is annotated as one of seven different emotions: anger, disgust, sadness, happiness, neutral, surprise, and fear. Additionally, MELD provides emotion labels for each utterance, including positive, negative, and neutral.

2.1.6   Conclusion

Table 2.1 demonstrates the richness and diversity of currently available multimodal sentiment analysis datasets, with the speaker language not limited to English. In recent years, datasets have been larger and of higher quality than those in the past, but two issues still exist with the current datasets. Firstly, there is currently no Chinese multimodal sentiment analysis dataset, which is not conducive to the development of Chinese sentiment analysis research. Secondly, although the label categories in the above datasets cover multimodal emotions, emotions, and attribute categories, these labels are only applicable to multimodal content and do not have any emotion or attribute labels in the single-modal dimension. In order to support the research work of this book, a Chinese dataset with unimodal emotion labeling and unified multimodal emotion labeling will be introduced in subsequent research to make up for the shortcomings of the existing datasets.

2.2 Multimodal Sentiment Analysis Dataset with Multilabel

2.2.1 Introduction

As a popular research direction in the field of natural language processing, sentiment analysis has attracted widespread attention. Previous studies mainly focused on text sentiment analysis [7, 8], inferring the speaker's emotional state through information such as vocabulary, syntax, and semantics in the text. These studies have achieved remarkable results, but they also have some limitations. First, the expression of text data is limited and cannot fully reflect the speaker's true emotional state, because the speaker's emotions may also be influenced by other factors. Second, text data is also easily affected by language ambiguity and misleading information, so it cannot accurately judge the speaker's emotional state. To address these issues, in recent years, researchers have begun to explore the application of nonverbal behavior in sentiment analysis [9, 10].

In multimodal sentiment analysis, there are two very important and challenging subtasks [11, 12], namely, unimodal representation and cross-modal fusion. Unimodal representation requires considering the temporal or spatial characteristics of different modalities, and convolutional neural networks (CNNs), long short-term memory (LSTM) networks, and deep neural networks (DNNs) are three commonly used methods for extracting unimodal features [13–15]. As for cross-modal fusion, many methods have been proposed recently, such as concatenation [13], tensor fusion network (TFN) [14], low-rank multimodal fusion (LMF) [16], memory fusion network (MFN) [15], dynamic fusion graph (DFG) [3], etc. In this section, we mainly focus on late fusion methods, which first learn unimodal representations and then perform cross-modal fusion. An intuitive idea is that the greater the differences between cross-modal representations, the better the complementarity of cross-modal fusion. However, existing late fusion models find it difficult to learn the differences between different modalities, which further limits the performance of fusion. The reason is that existing multimodal sentiment datasets only contain uniform multimodal annotations for each multimodal segment, which are not always applicable to all modalities. In other words, during the process of learning unimodal representations, all modalities share a standard annotation, and these uniform supervisions lead to more consistent unimodal representations, making it more difficult to distinguish between different modalities.

In this section, we present a new Chinese multimodal sentiment analysis dataset, named CH-SIMS. Unlike other existing multimodal datasets, our dataset not only contains multimodal annotations but also includes independent unimodal annotations (Fig. 2.1). This means that for each multimodal segment, our dataset provides more detailed and comprehensive annotations. Among them, M represents multimodal, T represents text, A represents audio, and V represents visual. The CH-SIMS dataset contains 2281 video segments from different movies, TV shows, and variety shows, which are finely annotated, including spontaneous expressions, different head poses, occlusions, and lighting conditions. Compared with other Chinese multimodal datasets, such as CHEAVD [17], our dataset not only contains three modalities but also provides independent unimodal annotations that can be used for both unimodal and multimodal sentiment analysis tasks. This provides researchers with a more comprehensive data resource for developing new multimodal sentiment analysis methods.

Fig. 2.1 An example of the annotation difference between CH-SIMS and other datasets

We propose a multimodal multitask learning framework that is based on SIMS and uses annotations from both single and multiple modalities. This framework allows for sharing of the underlying feature representation subnetwork between single-modal and multimodal tasks, and can be applied to any multimodal model based on postfusion. We introduce three postfusion models (TFN, LMF, and LF-DNN) into our framework and significantly improve the performance of multi-modal tasks through single-modal tasks. We also discuss in detail multimodal sentiment analysis, single-modal sentiment analysis, and multitask learning, and verify that the introduction of single-modal annotations can effectively expand the differences between different modalities, leading to better performance in modality fusion.

2.2.2 CH-SIMS Dataset

In this section, we introduced a novel Chinese sentiment analysis dataset, which is called CH-SIMS with independent single-modal annotation. Next, we will explain the process of data collection, annotation, and feature extraction in detail in sub-sections, and provide a detailed description and statistical information of the dataset.

Data Collection

Compared to a single-modal dataset, multimodal datasets have more stringent requirements. One basic requirement is that the speaker's face and voice must appear simultaneously in the video and be maintained for a certain period of time. In this section, in order to obtain video clips that are more close to real-life situations, we collected target segments from movies, TV dramas, and variety shows. After obtaining the raw videos, we used the video editing software Adobe Premiere Pro to perform frame-level editing of the target segments, which is a very time-consuming process but ensures the accuracy of the editing. In addition, we enforced the following constraints during the data collection and editing process:

1. We only focus on content spoken in Mandarin Chinese, and we are very cautious in selecting materials with accents.
2. The length of each clip should be between 1 s and 1 s.
3. In each video clip, no other person's face except for the speaker's face will appear.

Finally, we collected 60 raw videos and extracted 2281 video clips from them. SIMS has a rich variety of character backgrounds, a wide age range, and high quality. The basic statistical information can be found in Table 2.3.

Table 2.3 Statistics of SIMS dataset

Item	#
Total number of videos	60
Total number of segments	2281
Male	1500
Female	781
Total number of distinct speakers	474
Average length of segments	3.67
Average word count per segments	15

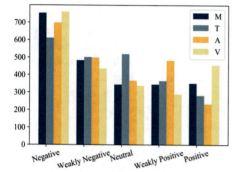
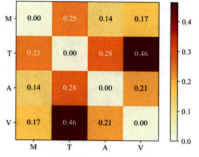

Fig. 2.2 The distribution of sentiment over the entire dataset in four annotations

Annotation

For each video segment, we need to perform multimodal annotation and three rounds of unimodal annotation. However, due to potential interference between different modalities that may cause confusion, we require annotators to only view information from the current modality during annotation to minimize this issue. Additionally, we do not allow annotators to perform all four annotations simultaneously. Specifically, annotators need to perform unimodal annotation first, followed by multimodal annotation. In order, the order of unimodal annotation is text, audio, and silent video, while multimodal annotation is performed last.

Each video clip is annotated by five students with a sentiment state of -1 (negative), 0 (neutral), or 1 (positive). To perform regression and multiclassification tasks, we average the five annotations and obtain a final sentiment label that is one of the values from $\{-1.0, -0.8, -0.6, -0.4, -0.2, 0.0, 0.2, 0.4, 0.6, 0.8, 1.0\}$. We further divide these values into 5 categories: negative $\{-1.0, -0.8\}$, weakly negative $\{-0.6, -0.4, -0.2\}$, neutral $\{0.0\}$, weakly positive $\{0.2, 0.4, 0.6\}$, and positive $\{0.8, 1.0\}$.

In Fig. 2.2, we can see the distribution of sentiment across the entire dataset. The left part displays the proportion of sentiment distribution in multimodal and three single-modal annotations, while the right part shows the annotation differences between different modalities. The larger the value, the greater is the difference. By observing the histogram and confusion matrix, we can find that there are more negative sentiment segments than positive ones, mainly because actors in film and television dramas are more expressive in negative sentiments than positive ones. The annotation difference in the confusion matrix is calculated by a specific formula:

$$D_{ij} = \frac{1}{N} \sum_{i=1}^{N} \left(A_i^n - A_j^n \right)^2 \tag{2.1}$$

where $i, j \in \{m, t, a, v\}$, N is the total number of samples, and each sample has a corresponding label value sequence, where A_i^n represents the nth label value of the sample in modality i.

According to the confusion matrix results, we can observe that the difference between A and M is very small, while the difference between V and T is very significant, which is consistent with our expectations. This is because the audio itself contains textual information, which is more in line with the concept of multimodality, while the connection between video and text is relatively sparse.

In addition to emotional annotations, our dataset also includes annotations for other attributes. However, in this experiment, we will primarily focus on emotional annotations as emotions are the main focus of our research.

Extracted Features

We introduce the process of feature extraction, which includes three modalities: text, audio, and visual. In all experiments, we use the same base features for analysis.

For the text modality, we manually transcribe all videos and only use Chinese transcriptions. Each transcription is augmented with two unique markers indicating the start and end. Then, we use pretrained Chinese BERT base word embeddings to obtain word vectors from the transcription [18]. It is worth noting that due to the nature of BERT, we do not use a segmentation tool as BERT can handle contextual information in language automatically. Finally, each word is represented as a 768-dimensional word vector.

For the audio modality, we use default parameters of the LibROSA [19] speech toolkit to extract acoustic features at 22050 Hz. We extract a total of 33-dimensional frame-level acoustic features, including 1-dimensional log F0, 20-dimensional Mel frequency cepstral coefficients (MFCCs), and 12-dimensional constant-Q chromagram (CQT). According to [20], these features are related to emotions and speech prosody and can be used for emotion analysis.

For the visual modality, we extract frames from video clips at a rate of 30 Hz. We use the MTCNN face detection algorithm [21] to extract aligned faces, we use the MultiComp OpenFace2.0 toolkit [11] to extract a set of 68 facial landmarks, 17 facial action units, head pose, head direction, and eye gaze. Finally, we extract a total of 709-dimensional frame-level visual features that can be used for emotion and expression recognition.

2.2.3   *Multimodal Multitask Learning Framework*

In this section, we will introduce a novel multimodal multitask learning framework proposed by us. Based on a multimodal learning framework called delayed fusion [13, 14], we added independent output units for three unimodal representations (text, audio, and visual), as shown in Fig. 2.3. The advantage of this framework is that each unimodal representation can not only participate in feature fusion but also be used to generate their predictive outputs, thereby improving the model's prediction accuracy. Specifically, our framework can perform multiple tasks simultaneously, such as image classification, speech recognition, and natural language processing. By organically integrating information from different modalities, our framework can fully leverage the advantages of each modality and ultimately achieve better performance.

In order to provide a clearer explanation of our method, we used three different modalities, namely, text, audio, and video, denoted by symbols t, a, and v, respectively. For the sake of convenience, we define L^u as the sequence length, D_i^u as the initial feature dimension, and D_r^u as the representation dimension learned by the unimodal feature extractors, where $u \in \{t, a, v\}$. In our method, we use a batch size of B. These symbols and definitions will be extensively used in the subsequent sections to ensure the clarity and accuracy of our method and results.

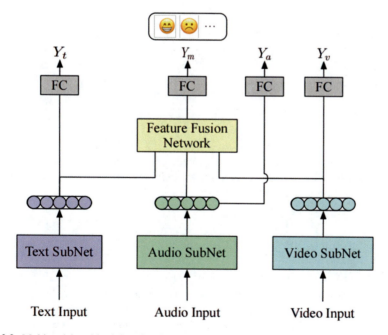

Fig. 2.3 Multimodal multitask learning framework

Unimodal Subnets

The purpose of a unimodal subnetwork is to learn a representation of a single modality from the raw feature sequence. The general feature extractor can be formalized as follows:

$$R_u = S_u\left(I_u\right) \tag{2.2}$$

where $I_u \in R^{B \times L^u \times D_i^u}$, $R_u \in R^{B \times D_r^u}$, $S_u\left(\cdot\right)$ is the feature extractor network for modal u.

We employed the approach described in references [14, 16] and utilized a three-layer Long Short-Term Memory (LSTM) [22] network with weight W_a and a three-layer deep neural network with weight W_v to extract embeddings for text, audio, and visual modalities in this work.

Feature Fusion Network

The purpose of the feature fusion network is to learn the cross-modal representation of three single-modal representations. The formalized expression of this representation is as follows:

$$R_m = F[(R_t, R_a, R_v)] \tag{2.3}$$

where $R_t, R_a, R_v \in R^{B \times D_r^u}$ are the unimodal representations. $F\left(\cdot\right)$ is the feature fusion network and R_m is the fusion representation.

The purpose of this section is to comprehensively compare three fusion methods: LF-DNN, TFN [14], and LMF [16], in order to contrast them with existing work.

Optimization Objectives

Our optimization goal is to select features within each modality by sparsely sharing parameters using L2 norm, in addition to training losses from different tasks. Therefore, our optimization objectives is:

$$\min \frac{1}{N_t} \sum_{n=1}^{N_t} \sum_i \alpha_i L(y_i^n, \widehat{y}_i^n) + \sum_j \beta_j \|W_j\|_2^2 \tag{2.4}$$

where $i \in \{m, t, a, v\}$, $j \in \{t, a, v\}$. nth represents the number of training samples. $L(y_i^n, \widehat{y}_i^n)$ represents the training loss of the nth sample in modality i. The parameter W_j is shared in modality j and multimodal tasks. The hyperparameter α_i is used to balance the effects of different tasks, while β_j represents the weight decay step size of sub-network j.

Finally, we adopted a three-layer deep neural network to generate results for various tasks. In this section, we treated these tasks as regression models and used the L1 loss from Eq. (2.4) as the training loss.

2.2.4   Experiments

This section will use SIMS (multimodal sentiment analysis system) to explore the following issues:

Multimodal Sentiment Analysis: In terms of multimodal sentiment analysis, we aim to verify the advantages of using single-modality annotations for multitask learning and establish a multimodal baseline for SIMS. We will compare the performance of different methods for multimodal sentiment analysis.

Unimodal Sentiment Analysis: In terms of single-modality sentiment analysis, we aim to verify the necessity of multimodal analysis and establish a single-modality baseline for SIMS. We will compare the performance of single-modality tasks using different annotation methods to determine which annotation method has a greater impact on single-modality sentiment analysis.

Representations Differences: Finally, we will compare the differences in single-modality representations between models with independent single-modality annotations and models with no annotations. t-SNE is a visualization technique that can project high-dimensional data into two-dimensional or three-dimensional space for visualization. Through this method, we will be able to better understand how SIMS learns more distinctive single-modality representations after using single-modality annotations.

Baselines

This section briefly introduces the baseline models used in the conducted experiments, which are primarily used for the fusion and processing of multimodal data. The following are detailed descriptions of these baseline models:

Early Fusion LSTM [23]: This model first concatenates the inputs from three modalities and then uses LSTM to capture long-distance dependencies in the sequence. This model aims to combine the information from multiple modalities early on and then process them to achieve interaction between modalities.

Later Fusion DNN: In contrast to EF-LSTM, this model learns single-modality features first and then concatenates them before classification. The main idea of this model is to process the information from each modality first and then combine them to achieve interaction between modalities.

Memory Fusion Network [15]: This model considers view-specific and cross-view interactions and models them continuously over time using a special attention mechanism, summarizing them over time through multiview gated memory.

MFN requires word-level alignment of three modalities, but since reliable Chinese corpus alignment tools are lacking, CTC [24] is used as an alternative method in this experiment.

Low-rank Multimodal Fusion [16]: This model performs effective multimodal fusion by using modality-specific low-rank factors to learn modality-specific and cross-modality interactions. The goal of this model is to reduce redundancy between modalities by efficiently using modality-specific low-rank factors.

Tensor Fusion Network [14]: Tensor Fusion Network (TFN): This model explicitly models view-specific and cross-view dynamics by creating a multidimensional tensor that spans single-modality, double-modality, and triple-modality interactions. The main idea of this model is to combine information from different modalities and represent it as a multidimensional tensor, and then use tensor decomposition to learn interaction between modalities.

Multimodal Transformer [25]: This model uses directional pairwise cross-modality attention to achieve interaction between multimodal sequences across different time steps and potentially adapt one modality to the flow of another. The main idea of this model is to use multihead self-attention mechanisms to process information from different modalities and use cross-modality attention mechanisms to achieve interaction between modalities.

Experimental Details

This section provides a detailed description of our experimental setup, including dataset partitioning, hyper-parameter selection, and evaluation metrics.

Dataset Splits: We first randomly shuffle all video clips and then divide them into training, validation, and testing sets based on multimodal annotations. The detailed partitioning results are shown in Table 2.4. We allocate the training, validation, and testing sets in a 6:2:2 ratio, where the labels NG: negative, WN: slightly negative, NU: neutral, WP: slightly positive, and PS: positive are used.

Hyper-parameters Selection: We first fix a specific sequence length for each modality. We select the maximum sequence length by using the average length plus three times the standard deviation based on empirical knowledge. In addition, we use binary classification accuracy to perform a grid search on all baselines and our method to adjust hyper-parameters. To ensure fair comparison, we use five random seeds (1, 12, 123, 1234, and 12,345) and record the average of five experimental results.

Table 2.4 Dataset splits in SIMS

Item	Total	NG	WN	NU	WP	PS
#Train	1368	452	290	207	208	211
#Valid	456	151	97	69	69	70
#Test	457	151	97	69	69	71

Evaluation Metrics: Similar to [3, 16], we record two types of experimental results: multiclass classification and regression. For multiclass classification, we record weighted F1 score and multiclass accuracy Acc-k, where $\in \{2, 3, 5\}$. For regression, we report mean absolute error (MAE) and Pearson correlation coefficient (Corr). A lower MAE indicates better performance, while higher values for other metrics indicate better performance.

Results and Discussion

This section will elaborate and discuss the experimental results of the research questions.

1. **Comparison with Baselines:** We compare three new multitask methods with a baseline method, focusing solely on multimodal evaluation results. Although these new methods are multitask, we only consider their performance in multimodal tasks. We list these results in Table 2.5. Compared with single-task models, we found that multitask models performed better in most evaluation metrics. In particular, all three improved models (MLF-DNN, MLFM, and MTFN) outperformed their corresponding original models (LF-DNN, LFM, and TFN) in all evaluation metrics except Acc-5. These results suggest that introducing independent unimodal annotations can significantly improve the

Table 2.5 (%) Results for sentiment analysis on the CH-SIMS dataset

Model	Acc-2	Acc-3	Acc-5	F1	MAE	Corr
EF-LSTM	69.37 ± 0.0	51.73 ± 2.0	21.02 ± 0.2	81.91 ± 0.0	59.34 ± 0.3	−04.39 ± 2.8
MFN	77.86 ± 0.4	63.89 ± 1.9	39.39 ± 1.8	78.22 ± 0.4	45.19 ± 1.2	55.18 ± 2.0
MULT	77.94 ± 0.9	65.03 ± 2.1	35.34 ± 2.9	79.10 ± 0.9	48.45 ± 2.6	55.94 ± 0.6
LF-DNN	79.87 ± 0.6	66.91 ± 1.2	41.62 ± 1.4	80.20 ± 0.6	42.01 ± 0.9	61.23 ± 1.8
MLF-DNN*	82.28 ± 1.3	69.06 ± 3.1	38.03 ± 6.0	82.52 ± 1.3	40.64 ± 2.0	67.47 ± 1.8
$\triangledown$	↑2.41	↑2.15	↓3.59	↑2.32	↓1.37	↑6.24
LMF	79.34 ± 0.4	64.38 ± 2.1	35.14 ± 4.6	79.96 ± 0.6	43.99 ± 1.6	60.00 ± 1.3
MLMF*	82.32 ± 0.5	67.70 ± 2.2	37.33 ± 2.5	82.66 ± 0.7	42.03 ± 0.9	63.13 ± 1.9
$\triangledown$	↑2.98	↑3.32	↑2.19	↑2.70	↓1.96	↑3.13
TFN	80.66 ± 1.4	64.46 ± 1.7	38.38 ± 3.6	81.62 ± 1.1	42.52 ± 1.1	61.18 ± 1.2
MTFN*	82.45 ± 1.3	69.02 ± 0.3	37.20 ± 1.8	82.56 ± 1.2	40.66 ± 1.1	66.98 ± 1.3
$\triangledown$	↑1.79	↑4.56	↓1.18	↑0.94	↓1.86	↑5.80

performance of existing methods in multimodal sentiment analysis. Furthermore, we found that some methods performed well on existing public datasets but poorly on our proposed SIMS dataset. This further underscores the challenging task of designing a powerful, cross-lingual multimodal sentiment analysis model. It is for this reason that we propose this new dataset to promote research and further development in this field.

2. **Unimodal Sentiment Analysis:** The existence of independent unimodal annotations in SIMS requires us to explore the ability of unimodal sentiment analysis. Therefore, we conducted two sets of experiments on unimodal sentiment analysis and the results are listed in Table 2.6. In the first set of experiments, we used real unimodal labels to verify the model's ability to perform unimodal sentiment analysis. The results showed that the performance using unimodal labels was better than using multimodal labels in the same unimodal task. This means that the model performs well in unimodal sentiment analysis, but these results cannot accurately reflect the speaker's actual emotional state. To address this issue, we conducted a second set of experiments to verify the model's ability to predict the speaker's true emotions using only unimodal information by using multimodal labels. The results showed that the performance of using only unimodal information was lower than that of conducting sentiment analysis using multimodal information. This indicates that using only unimodal information for sentiment analysis is insufficient due to the inherent limitations of unimodal information.

3. **Representations Differences:** As shown in Fig. 2.4, we visualized the differences between the learned single-modality representations of different models using t-SNE [26] algorithm, and the results showed that the new models (MLF-DNN, MTFN, and MLMF) could better distinguish the differences between different single-modalities compared to the original models (LF-DNN, TFN, and LMF). This indicates that single-modality annotations can provide richer information for the models, thus improving the performance of multimodal tasks. In addition, by improving the complementarity between modalities, single-modality annotations can promote effective interaction between different modalities, thus better achieving the goals of multimodal tasks.

2.2.5 Conclusion

The novel multimodal emotion analysis dataset CH-SIMS proposed in this section contains independent unimodal annotations, which provide more accurate emotional information for model learning. Regarding the multimodal multitask learning framework, we designed it based on the postfusion method to enhance complementarity and learning effectiveness among modalities. Through numerous experiments, we found that a unified multimodal annotation may not always reflect the emotional states of individual modalities, and unimodal annotations can help models learn more differentiated information. In multitask learning, asynchronous learning of

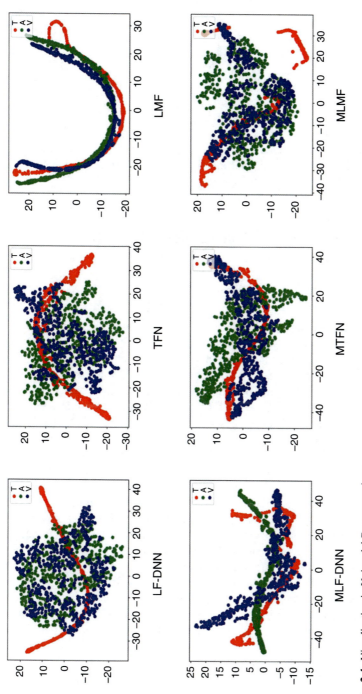

Fig. 2.4 Visualization in Unimodal Representations

Table 2.6 (%) Results for unimodal sentiment analysis on the CH-SIMS dataset using MLF-DNN

Task	Label	Acc-2	F1	MAE	Corr
A	V	67.70	79.61	53.80	10.07
	M	65.47	71.44	57.89	14.54
V	V	81.62	82.73	49.57	57.61
	M	74.44	79.55	54.46	38.76
T	T	80.26	82.93	41.79	49.33
	M	75.19	78.43	52.73	38.55

different sub-tasks may have adverse effects on multimodal emotion analysis. In future research, we will continue to explore the relationship between multimodal analysis and multitask learning and attempt more fusion strategies to improve model performance and generalization ability.

2.3   An Extension and Enhancement of the CH-SIMS Dataset

2.3.1   Introduction

This section discusses the limitations of existing multimodal sentiment analysis (MSA) corpora, including the blurriness and flat color tones of videos, as well as the triviality of single-modality solutions resulting from label bias. To overcome these challenges, a high-quality MSA dataset is needed, which should have a certain scale, be composed of diverse nonlinguistic contextual information, have a balanced instance annotation distribution, and include single-modality annotations.

To address these challenges, researchers have constructed the CH-SIMS v2.0 dataset, an enhanced and expanded version of the CH-SIMS (Chinese multimodal sentiment dataset), which includes 4402 supervised data with single-modality annotations and over 10,000 unsupervised data. The dataset was collected from 11 different scenarios to simulate real human-machine interaction scenes, paving the way for emotion computing applications (ECA) to understand user emotions.

As shown in Fig. 2.5, the multimodal annotations in the CH-SIMS v2.0 dataset use colors from red (strongly negative) to green (strongly positive) to reflect emotional intensity, while the colors of the single-modality annotations on the right reflect the consistency between single-modality and multimodal emotions [27]. Instances marked with "*" are unsupervised data. Through this dataset, researchers can conduct better acoustic and visual representation learning and perform fine-grained sentiment intensity prediction, thus improving the accuracy and effectiveness of multimodal sentiment analysis.

Fig. 2.5 Illustration of the constructed CH-SIMS v2.0 dataset

2.3.2   CH-SIMS V2.0 Dataset

Data Collection

Some constraints were followed while collecting the raw video collection for this job. Specifically, to meet the requirements of the visual modality, the collected videos had to be of high definition and have the appearance of a speaker. For the acoustic modality, the videos had to be in Mandarin, to ensure consistency of the speech data. As for the textual modality, accurate transcription was required, which involved a certain amount of manpower and time cost. Throughout the process, the raw video clips were kept in their original resolution and recorded in MP4 format. These raw video clips are the basic dataset, which need to be further processed to obtain usable data. To obtain high-quality instances, the popular video editing tool "PlotPlayer" was used to crop the videos at the frame level, in order to better meet the needs of the training model. These improvements are reflected in two aspects, compared to the previous CH-SIMS dataset.

Diverse video scenarios: Video clips from different scenarios may have different emotional tendencies. Traditional face detection tools, such as MTCNN [28], have some defects and cannot distinguish speakers' faces in multiparty scenarios, so the previously CH-SIMS dataset [27] had limited scene resources. To address this issue, this section uses TalkNet [29] as an Active Speaker Detection (ASD) tool to collect video scenes containing multiple faces. To enrich the diversity of the scenes, the CH-SIMS v2.0 dataset collects data from various scenarios such as emotional drama, interviews, modern television, talk shows, vlogs, movies, period dramas, variety shows, and other raw video clips. Simulate the complexity of real-world scenarios, there may be variations in the angles, distances, and lighting conditions between different videos. In this dataset, instances lacking speakers' faces will be removed to improve the quality of instances. At the same time, instances with front, side, and oblique faces will also be collected to ensure the diversity of the dataset. In addition, instances' acoustics may contain slight noise, such as environmental music and white noise, and speakers' tone and speed may also vary. These details are all to better simulate real-world scenarios and make the dataset more diverse and comprehensive.

Expressive acoustic and visual behavior: The proposed dataset focuses more on nonverbal behaviors compared to CH-SIMS, as nonverbal behaviors can convey the speaker's emotions and intentions more intuitively. However, not all nonverbal behaviors in the raw clipped videos necessarily carry emotional expression, so when selecting data, nonverbal behaviors without emotional expression were intentionally excluded. For example, in Fig. 2.5, the neutral textual modality "What does he like about you?" needs to capture the smiling facial expression from the visual modality to better express the speaker's emotions. On the other hand, in the instance "I really like and love this boy," the textual modality can mislead the model's prediction of the speaker's emotions because the sentence appears positive on the surface, but there may be other emotional elements such as the speaker's doubts, anxieties, or

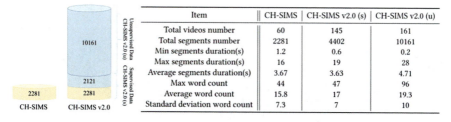

Item	CH-SIMS	CH-SIMS v2.0 (s)	CH-SIMS v2.0 (u)
Total videos number	60	145	161
Total segments number	2281	4402	10161
Min segments duration(s)	1.2	0.6	0.2
Max segments duration(s)	16	19	28
Average segments duration(s)	3.67	3.63	4.71
Max word count	44	47	96
Average word count	15.8	17	19.3
Standard deviation word count	7.3	7	10

Fig. 2.6   Statistics of each component of the CH-SIMS v2.0 dataset

other negative emotions. In addition, there are other instances with ambiguous, ironic, and metaphorical textual modalities that require more accurate emotional expression from nonverbal behaviors. Therefore, the CH-SIMS v2.0 dataset contains a large number of instances that depend on weak textual modalities, which also means that models based on text or text-based modalities may not accurately predict the speaker's emotions. Therefore, it is necessary to consider multiple sources of information to improve the accuracy of emotion recognition.

The final statistical data of CH-SIMS v2.0 is shown in Fig. 2.6. The CH-SIMS v2.0 dataset consists of many supervised and unsupervised instances. Among them, there are 4402 supervised instances represented as CH-SIMS v2.0(s), and the annotation information of these instances is known in the dataset. In contrast, there are 10,161 unsupervised instances represented as CH-SIMS v2.0(u), and the annotation information of these instances is unknown, requiring training with methods such as self-supervised learning. In addition, CH-SIMS v2.0(s) has similar properties to the original CH-SIMS dataset, so it can be used for supervised learning tasks' training and evaluation. CH-SIMS v2.0(u) has a more diverse duration distribution, which enables the dataset to better simulate situations in the real world and improve the model's generalization ability. It should be noted that since the text modality of unsupervised instances is directly extracted from text records without manual processing, there may be some noise and inaccurate information. To obtain better performance when using the CH-SIMS v2.0 dataset, methods are needed to reduce the impact of this noise.

Data Annotation

When annotating the supervised CH-SIMS v2.0 (s) dataset, it is necessary to collect raw video segments. This dataset is consistent with the previous CH-SIMS dataset, where each instance is annotated with fine-grained sentiment unimodal and multi-modal labels ranging from Strong Negative (−3), Negative (−2), Weak Negative (−1), Neutral (0), Weak Positive (1), Positive (2), to Strong Positive (3). To improve fairness, several instances of each sentiment class are provided to annotators as standards before annotation. This ensures that annotators can capture emotional differences and nuances as accurately as possible when annotating the data. Seven annotators participate in the data annotation process. To improve the quality of

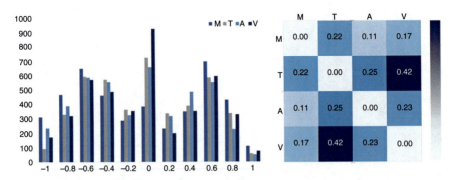

Fig. 2.7 Left: the distribution of sentiment over the entire dataset in one Multimodal annotation and three unimodal (Text, Acoustic, and Visual) annotations. Right: the confusion matrix shows the annotations difference between different modalities in CH-SIMS (s) v2.0. The larger the value, the greater the difference

annotation, quality control is needed. Specifically, in the postprocessing stage, the highest and lowest scores are removed, and the average score of the remaining five results is mapped to the space of Strong Negative $(-1, -0.8)$, Weak Negative $(-0.6, -0.4, -0.2)$, Neutral (0), Weak Positive $(0.2, 0.4, 0.6)$, and Strong Positive $(0.8, 1.0)$ as the final sentiment label. This method ensures the consistency and accuracy of the data, thereby improving the performance of the model.

CH-SIMS v2.0 is a dataset used for emotion analysis and recognition tasks of multimodal signals, including speech, vision, and text. In this version, a strict modality isolation strategy is adopted to improve the single-modal label annotation process and enhance the annotation quality of the dataset. Specifically, CH-SIMS v2.0 improves the accuracy and consistency of the dataset by re-annotating the original CH-SIMS dataset's single-modal annotations. Regarding acoustic modality annotation, CH-SIMS v2.0 injects noise to blur spoken words, ensuring the consistency of overall rhythm and acoustic features. For visual modality annotation, videos are presented in mute mode, and text descriptions are removed to eliminate the impact of interactive mode information on annotators. For text modality annotation, only textual descriptions are provided to keep annotators focused on the emotional content of the text itself. Moreover, to reduce the impact of interactive mode information on annotators, CH-SIMS v2.0 adopts the annotation order of text, acoustic, visual, and multimodal. This annotation order ensures that the annotation of each modality is independent of other modalities, thereby improving the consistency and credibility of the annotations.

The statistical data mentioned in this article is shown on the left side of Fig. 2.7. Through the data, it can be observed that in the CH-SIMS v2.0 dataset, the ratio of positive and negative instances is 83.41%, which is higher than the original CH-SIMS dataset of 56.38%, indicating more emotion labels in the dataset. At the same time, most instances have weak emotion intensity, indicating that the emotion labels in the dataset tend to be neutral or weak. In addition, by comparing the number of neutral instances in different single-modality and multimodal annotations, it can be concluded that multimodal annotations contain more emotions than each single modality, especially the visual modality showing the importance of integrating multimodal information. This suggests that better emotion capture can be achieved by combining information from multiple modalities. Reference [27] mentioned the average difference between each single-modality and multimodal label, which is explained on the right side of Fig. 2.7. The results show a significant difference between single-modality and multimodal labels, validating the assumption that a unified multimodal label cannot reflect single-modality emotions. This also suggests that designing independent emotion labels for different modalities is necessary. Finally, it is found that the acoustic and multimodal annotations are most similar, while the text annotation is the least similar. This means that using nonverbal cues can improve the performance of emotion prediction because the emotional information in different modalities is different. Therefore, in order to predict emotions more accurately, all available information should be utilized and integrated.

2.3.3   Feature Extraction

The default modal sequence of CH-SIMS v2.0 is extracted by an open-source integrated feature extraction tool called MMSA-FET [29]. The MMSA-FET tool can analyze and process the dataset to extract key feature information. When conducting experiments using CH-SIMS v2.0, the feature sequence extracted by MMSA-FET will be used by default if no additional description is provided.

Default Textual Feature: BERT [18] is currently one of the most popular models in natural language processing and has shown outstanding performance in many tasks. We used the pretrained BERT model Bert-base-Chinese to learn contextual word embeddings as effective textual features. By inputting the text sequence into the BERT model, we can obtain the semantic representation of each word in context, which is called word embedding. To make the input sequence length consistent, we fixed the length of the token sequence to 50 and can use padding or truncation methods. After the input sequence passes through the BERT model, we can obtain a 50×768 matrix, where each row represents the word embedding of a single word. To obtain the feature representation of the entire text, we add each row vector in this matrix, ultimately obtaining a 768-dimensional textual feature vector. Using the pretrained BERT model for text feature extraction can reduce reliance on specific domains and improve model generalization. At the same time, BERT's excellent performance has made it a preferred model for many natural language processing tasks.

Default Acoustic Feature: To obtain more accurate results, it is necessary to extract representative acoustic features, among which Low Level Descriptors (LLD) are a common type of acoustic feature. We used a sampling rate of 16,000 Hz for digital processing of sound signals, and extracted 25-dimensional LLD features of GeMAPS [30] using the OpenSMILE [31] backend. To ensure that audio signals of different lengths can be processed, the final acoustic features were padded or truncated to a sequence length of 925.

Default Visual Feature: In multimodal emotion recognition, visual modality is a very important feature type. To obtain effective visual features, frame extraction from the video is needed. We use FFmpeg tool to extract images from the video at a rate of 25 frames per second. Then, an effective Action Speaker Detection (ASD) method, TalkNet [32],, is used to detect the speaker's face from all faces in each image. When ASD fails to detect the speaker's face, the image is dropped, and instances with over 25% missing images are discarded. After ASD, we use the OpenFace [31] backend to extract facial features, including 68 facial landmarks, 17 facial action units, head pose, head orientation, and eye gaze direction. Finally, the 177-dimensional frame-level visual features are padded or truncated to a sequence length of 232 for further processing and analysis.

2.3.4 Acoustic Visual Mixup Consistent (AV-MC) Framework

Although the CH-SIMS v2.0 dataset provides as many multimodal scenes as possible, these scenes still cannot cover all multimodal contexts. To improve the performance of the MSA model, we have developed a new approach called Audio-Visual Mix Consistency (AV-MC) framework. The framework utilizes the similarity between sound and image to mix them together and create new samples. These new samples contain features from two different samples, which can provide more diversity and robustness to the model. At the same time, we introduce a consistency loss function to ensure that the mixed data still has the same emotional labels, thus avoiding the model associating emotional labels only with specific modalities. The overall structure of the AV-MC framework is shown in Fig. 2.8.

2.3.5 Experiments

Benchmark Results on CH-SIMS v2.0

A benchmark model refers to a benchmark system used to evaluate the performance of other models in multimodal learning. Benchmark models can be divided into two categories based on whether they use single-modality annotations. One category is the traditional MSA models for supervised learning using a unified multimodal annotation. These models learn multimodal representations by fusing multiple modalities, such as the late-fusion deep neural network (LF_DNN) [33], tensor fusion network (TFN) [34], low-rank multimodal fusion (LMF) [35], memory fusion network (MFN) [15], graph memory fusion network (Graph_MFN) [3], multimodal transformer (MulT) [25], Bert network's multimodal adaptation gate (Bert_MAG)

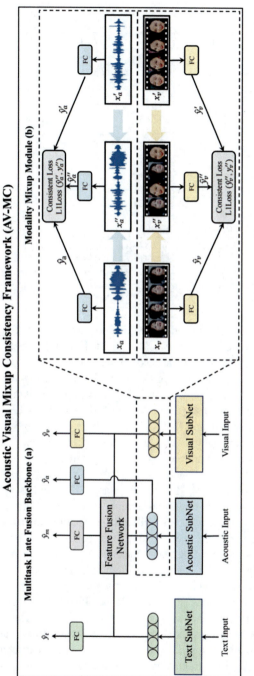

Fig. 2.8 Acoustic Visual Mixup Consistent (AV-MC) Framework under semisupervised learning paradigm, which consists of Multitask Late Fusion Backbone (**a**) with Modality Mixup Module (**b**)

Table 2.7 Multimodal sentiment analysis model on CH-SIMS v2.0 dataset

Models	Acc2 (↑)	F1_score (↑)	Acc2_weak (↑)	Corr (↑)	R_square (↑)	MAE (↓)
LF_DNN	73.95	73.84	69.13	52.19	20.84	0.381
TFN	76.51	76.31	66.27	66.65	35.90	0.323
LMF	77.05	77.02	69.34	63.75	40.64	0.343
MFN	75.27	75.24	66.46	60.60	32.26	0.355
Graph_MFN	73.98	73.62	69.82	49.71	13.78	0.396
MulT	79.50	79.59	69.61	70.32	47.15	0.317
Bert_MAG	79.79	79.78	71.87	69.09	43.08	0.334
MISA	80.53	80.63	70.50	72.49	50.59	0.314
MMIM	80.95	80.97	72.28	70.65	43.81	0.316
Self MM	79.01	78.89	71.87	64.03	29.36	0.335
MLF_DNN*	78.40	78.44	71.59	65.80	39.34	0.33
MTFN*	80.26	80.33	71.07	70.54	46.07	0.318
MLMF*	79.92	79.72	69.88	71.37	47.53	0.302
AV-MC*	82.50 (2.00%)	82.55 (2.01%)	74.54 (3.13%)	73.17 (0.94%)	50.65 (0.12%)	0.297 (1.66%)
AV-MC (Semi)*	83.46 (3.10%)	83.52 (3.15%)	74.54 (3.13%)	76.04 (4.90%)	57.37 (13.40%)	0.286 (5.30%)

[36], modality invariant and specific representation network (MISA) [37], multi-modal information maximization network (MMIM) [38], self-supervised multitask learning network (Self_MM) [39], and so on. The other category utilizes single-modality annotations to guide the learning of single-modality representations, and aims to learn single-modality representations. For example, multitask tensor fusion network (MTFN), multitask late-fusion deep neural network (MLF_DNN), multi-task low-rank multimodal fusion (MLMF), etc. These models train multiple tasks using single-modality annotations and share network parameters among tasks to achieve the goal of learning multiple tasks in a single model.

Table 2.7 shows the performance of various baseline models and the proposed AV-MC framework on the CH-SIMS v2.0 dataset. The models marked with an asterisk (*) were trained with multitask learning, and "Semi" indicates the use of additional unsupervised data. Across different evaluation metrics, all baseline models exhibited lower Acc2_weak and poorer performance on all regression metrics. This suggests that while these baseline models perform well in basic sentiment polarity classification, they fail in predicting fine-grained sentiment intensity. By comparing two different category baseline models, it was found that a model using single-modal annotations and a simple postfusion backbone network can achieve competitive performance with state-of-the-art models. The reliable performance of the second category baseline model further demonstrates the effectiveness of single-modal annotations in predicting fine-grained sentiment intensity. In addition, the proposed AV-MC framework outperforms all existing MSA baseline models, especially in distinguishing weakly sentimental instances under supervised conditions, validating the effectiveness of modality fusion strategy in predicting fine-grained sentiment intensity. Finally, using CH-SIMS v2.0 (u) can further improve model performance, revealing the potential improvement of using unsupervised data. This suggests that using more data can improve model performance in sentiment analysis tasks, as the model can learn more features and patterns from more data. This also demonstrates the helpfulness of unsupervised learning in sentiment analysis tasks.

Case Study

This section introduces an AV-MC framework that can handle multimodal data and make predictions on fine-grained emotions. Specifically, as shown in Fig. 2.9, the AV-MC framework records verbal language, nonverbal behavior, single-modal annotations, and multimodal annotations for each instance. Cases where there are inconsistencies between single-modal and multimodal annotations are marked in red. In addition, the article also provides the prediction results and MAE values of the AV-MC framework. The application of the AV-MC framework is demonstrated through typical cases of constructing CH-SIMS v2.0, and compared with corresponding predictions and annotations. In these cases, diverse nonverbal emotional behaviors can be observed, such as obvious anger, disappointment in tone, and smiling or stiff facial expressions. These nonverbal behaviors are crucial for fine-grained emotion prediction and can alleviate the problem of relying mainly on text. Meanwhile, the AV-MC framework performs well when the modalities are consistent. For example, in the case of #I-2, there is no obvious emotion expressed in the text modality, but the AV-MC framework can still make predictions through negative acoustic and visual cues. In the case of #I-5, despite the smiling face, the AV-MC still failed to predict positive emotions, because the majority of modalities (text and acoustic) expressed negative emotions. These results indicate that the AV-MC module can improve performance by providing better single-modal representation learning. However, due to the limitations of the simple late fusion strategy, the AV-MC framework performs poorly in emotion reasoning. This indicates that the AV-MC framework still needs improvement to better handle multimodal data and improve the accuracy of emotion prediction.

CH-SIMS v2.0 is a framework that can help achieve robust real-world scenario applications, providing rich resources for fully exploring nonverbal behaviors. In the future, MSA models need to pay more attention to the design of unimodal encoders to capture modality-specific features. At the same time, the design of the fusion module is also critical, because it can integrate unimodal cues into sentiment inference, thereby improving the performance and accuracy of the model.

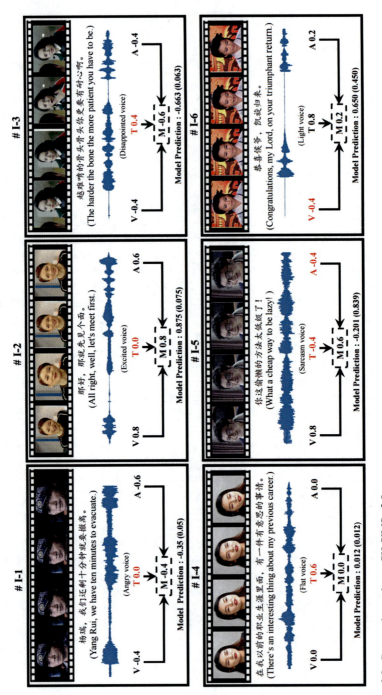

Fig. 2.9 Case study results on CH-SIMS v2.0

2.3.6 Conclusion

This section mainly introduces how to improve the contribution of nonverbal cues to sentiment analysis from both resource and method perspectives. In terms of resources, we propose a new semisupervised dataset, CH-SIMS v2.0, which includes videos with expressive nonverbal behaviors and provides fine-grained unimodal and multimodal annotations. By comparing the differences between textual and multi-modal annotations, we find that utilizing emotion-bearing acoustic and visual cues is crucial for predicting sentiment analysis. In terms of methods, we propose two new approaches to improve the performance of multimodal sentiment analysis. One approach is to design interpretable methods to discover decisive modalities based on unimodal annotations, which enhances the accuracy and interpretability of multimodal sentiment analysis. Another approach is the unsupervised data-based emotion-bearing multimodal pretraining method, which can learn rich emotion-bearing knowledge and improve the performance of multimodal sentiment analysis. In the future, we can explore more approaches to improve the contribution of nonverbal cues to sentiment analysis.

2.4 Summary

This chapter provides an overview of the main topics and objectives discussed in the book. It begins by highlighting the existing multimodal sentiment analysis datasets, showcasing their richness and diversity across different languages. However, two significant issues are identified. Firstly, there is a lack of Chinese multimodal sentiment analysis datasets, hindering research development in Chinese sentiment analysis. Secondly, the current datasets only provide multimodal emotion and attribute labels, lacking emotion or attribute labels in the single-modal dimension.

To address these shortcomings, this chapter introduces the CH-SIMS dataset, which includes independent unimodal annotations, providing more accurate emotional information for model learning. This chapter also presents a multimodal multitask learning framework based on postfusion methods, aiming to enhance complementarity and learning effectiveness among modalities. The importance of unimodal annotations in capturing differentiated information is emphasized, and the adverse effects of asynchronous learning in multitask settings are discussed.

Additionally, this chapter explores the improvement of nonverbal cues in sentiment analysis from resource and method perspectives. A new semisupervised dataset, CH-SIMS v2.0, is proposed, containing expressive nonverbal behaviors and providing fine-grained unimodal and multimodal annotations. The differences between textual and multimodal annotations highlight the significance of utilizing emotion-bearing acoustic and visual cues for sentiment analysis. Two novel approaches are suggested to enhance multimodal sentiment analysis: designing interpretable methods based on unimodal annotations to identify decisive modalities and employing unsupervised data-based emotion-bearing multimodal pretraining methods to improve performance.

References

1. Goodfellow I, Bengio Y, Courville A (2016) Deep learning. MIT Press, Cambridge, MA
2. Zadeh A, Zellers R, Pincus E, et al (2016) Mosi: multimodal corpus of sentiment intensity and subjectivity analysis in online opinion videos. arXiv preprint arXiv:1606.06259
3. Zadeh AAB, Liang PP, Poria S, et al (2018) Multimodal language analysis in the wild: Cmu-mosei dataset and interpretable dynamic fusion graph. Proceedings of the 56th Annual Meeting of the Association for Computational Linguistics, (1): 2236–2246
4. Busso C, Bulut M, Lee CC et al (2008) IEMOCAP: interactive emotional dyadic motion capture database. Lang Resour Eval 42(4):335–359
5. Poria S, Hazarika D, Majumder N, et al (2019) MELD: A Multimodal Multi-Party Dataset for Emotion Recognition in Conversation. Proceedings of the 57th Annual Meeting of the Association for Computational Linguistics, 527–536
6. Hsu CC, Chen SY, Kuo CC, et al (2018) Emotion lines: An emotion corpus of multi-party conversations. Proceedings of the 8th International Conference on Language Resources and Evaluation, 1597–1601
7. Pang B, Lee L (2008) Opinion mining and sentiment analysis. Found Trends Inf Retr 2(1–2): 1–135
8. Liu B, Zhang L (2012) A survey of opinion mining and sentiment analysis. Mining Text Data:415–463
9. Bing L (2012) Sentiment analysis and opinion mining. Synth Lect Hum Lang Technol 5(1): 1–167
10. Poria S, Cambria E, Hazarika D, Mazumder N, et al. (2017) Multi-level multiple at tentions for contextual multimodal sentiment analysis. Proceedings of the 2017 IEEE International Conference on Data Mining, 1033–1038
11. Baltrusaitis T, Ahuja C, Morency L-P (2019) Multimodal machine learning: a survey and taxonomy. IEEE Trans Pattern Anal Mach Intell 41(2):423–443
12. Guo W, Wang J, Wang S (2019) Deep multimodal representation learning: a survey. IEEE Access 7:63373–63394
13. Cambria E, Hazarika D, Poria S, et al. (2017) Benchmarking multimodal sentiment analysis. Proceedings of the 2017 International Conference on Computational Linguistics and Intelligent Text Processing, 166–179
14. Zadeh A, Chen M, Poria S, et al. (2017) Tensor fusion network for multimodal sentiment analysis. Proceedings of the 2017 Association for Computational Linguistics Conference on Empirical Methods in Natural Language Processing, 1103–1114
15. Zadeh A, Liang PP, Mazumder N, et al (2018) Memory fusion network for multi-view sequential learning. Proceedings of the 32nd Association for the Advancement of Artificial Intelligence Conference on Artificial Intelligence, 5634–5641
16. Liu Z, Shen Y, Lakshminarasimhan VB, et al (2018) Efficient low-rank multimodal fusion with modality-specific factors. Proceedings of the 56th Annual Meeting of the Association for Computational Linguistics, (1):2247–2256
17. Li Y, Tao J, Chao L, Bao W et al (2016) Cheavd: A Chinese natural emotional audio–visual database. J Ambient Intell Humaniz Comput 8(6):913–924
18. Devlin J, Chang MW, Lee K, et al. (2019) Bert: Pre-training of deep bidirectional transformers for language understanding. Proceedings of the 2019 Conference of the North American

Chapter of the Association for Computational Linguistics: Human Language Technologies, (1): 4171–4186

19. McFee B, Raffel C, Liang D, et al. (2015) Librosa: Audio and music signal analysis in python. Proceedings of the 14th python in science conference, 18–25

20. Li R, Wu Z, Jia J, et al. (2018) Inferring user emotive state changes in realistic human-computer conversational dialogs. Proceedings of the 26th Association for Computing Machinery international conference on Multimedia, 136–144

21. Zhang K, Zhang Z, Li Z et al (2016) Joint face detection and alignment using multitask cascaded convolutional networks. IEEE Sig Process Lett 23(10):1499–1503

22. Hochreiter S, Schmidhuber J (1997) Long short-term memory. Neural Comput 9(8):1735–1780

23. Williams J, Kleinegesse S, Comanescu R, et al (2018) Recognizing emotions in video using multimodal DNN feature fusion. Proceedings of Grand Challenge and Workshop on Human Multimodal Language, 11–19

24. Graves A, Fernandez S, Gomez F, et al (2006) Connectionist temporal classification: labelling unsegmented sequence data with recurrent neural networks. Proceedings of the 23rd international conference on Machine learning, 369–376

25. Tsai YHH, Bai S, Liang PP, et al (2019) Multimodal transformer for unaligned multimodal language sequences. Proceedings of the 57th Annual Meeting of the Association for Computational Linguistics, 6558–6569

26. van der Maaten L, Hinton G (2008) Visualizing data using t-sne. J Mach Learn Res 9(86): 2579–2605

27. Yu W, Xu H, Fanyang Meng, et al (2020) Ch-sims: A Chinese multimodal sentiment analysis dataset with fine-grained annotation of modality. Proceedings of the 58th Annual Meeting of the Association for Computational Linguistics. 3718–3727

28. Wöllmer M, Metallinou A, Eyben F, et al (2010) Context-sensitive multimodal emotion recognition from speech and facial expression using bidirectional LSTM modeling. Proceedings of InterSpeech, 2362–2365

29. Mao H, Yuan Z, Xu H, et al (2022) M-SENA: An integrated platform for multimodal sentiment analysis. Proceedings of the 60th Annual Meeting of the Association for Computational Linguistics: System Demonstrations. 204–213

30. Eyben F, Scherer KR, Schuller BW et al (2015) The Geneva minimalistic acoustic parameter set (GeMAPS) for voice research and affective computing. IEEE Trans Affect Comput 7(2): 190–202

31. Eyben F, Wöllmer M, Schuller B (2010) Opensmile: the munich versatile and fast open-source audio feature extractor. Proceedings of the 18th Association for Computing Machinery international conference on Multimedia, 1459–1462

32. Tao R, Pan Z, Das RK, et al (2021) Is someone speaking? Exploring long-term temporal features for audio-visual active speaker detection. Proceedings of the 29th Association for Computing Machinery International Conference on Multimedia. 3927–3935

33. Luo H, Ji L, Huang Y, Wang B, Ji S, Li T (2021) ScaleVLAD: improving multimodal sentiment analysis via multi-scale fusion of locally descriptors. arXiv preprint arXiv:2112.01368 ,2021

34. Zadeh A, Chen M, Poria S, Cambria E, Morency LP (2017) Tensor fusion network for multimodal sentiment analysis. arXiv preprint arXiv:1707.07250 ,2017

35. Liu Z, Shen Y, Lakshminarasimhan VB, Liang PP, Zadeh A, Morency LP (2018) Efficient low-rank multimodal fusion with modality-specific factors. arXiv preprint arXiv:1806.00064

36. Rahman W, Hasan MK, Lee S, et al (2020) Integrating multimodal information in large pretrained transformers. Proceedings of the 58th Annual Meeting of the Association for Computational Linguistics, 2359–2369

37. Hazarika D, Zimmermann R, Poria S (2020) Misa: Modality-invariant and-specific representations for multimodal sentiment analysis. Proceedings of the 28th Association for Computing Machinery International Conference on Multimedia, 1122–1131

38. Han W, Chen H, Poria S (2021) Improving multimodal fusion with hierarchical mutual information maximization for multimodal sentiment analysis. Proceedings of the 2021 Conference on Empirical Methods in Natural Language Processing, 9180–9192
39. Wenmeng Y, Hua X, Yuan Z et al (2021) Learning modality-specific representations with self-supervised multi-task learning for multimodal sentiment analysis. Proce Assoc Advanc Artific Intellig Conf Artific Intellig 35(12):10790–10797

Chapter 3
Early Unimodal Sentiment Analysis of Comment Text Based on Traditional Machine Learning

Abstract This chapter introduces the main topics and objectives discussed in subsequent sections, covering various aspects of opinion mining and sentiment analysis. It addresses different challenges and proposes novel methods.

Section 3.1 highlights the need to filter out irrelevant information and personal attacks in online discussions to focus on evaluative opinion sentences. It proposes an unsupervised method utilizing natural language processing techniques and machine learning algorithms to automatically filter and classify sentences with evaluative opinions. This section calls for further research to explore more precise and efficient methods for identifying evaluative opinions and their application in sentiment polarity analysis. Section 3.3 presents a novel subproblem in opinion mining, focusing on grouping feature expressions in product reviews. It argues for the necessity of user supervision in practical applications and proposes an EM formulation enhanced with soft constraints to achieve accurate opinion summaries. This section showcases the competence and generality of the proposed method through experimental results from various domains and languages. Section 3.1 introduces the use of topic modeling, specifically the LDA method, for sentiment mining. It extends the LDA method to handle large-scale constraints and proposes two methods for automatically extracting constraints to guide the topic modeling process. The constrained-LDA model and extracted constraints are then applied to group product features, demonstrating superior performance compared to other methods. Section 3.3 addresses the challenge of grouping synonyms in opinion mining, proposing an efficient method based on similarity measurement. Experimental results from different domains validate the effectiveness of the method. Section 3.3 focuses on feature extraction for sentiment classification and compares the impact of different types of features through experimental analysis. This section provides an in-depth study of all feature types and discusses key problems associated with feature extraction algorithms. Section 3.2 explores the use of unsupervised learning methods for sentiment classification, emphasizing their advantages in classifying opinionated texts at different levels and for feature-based opinion mining. This section presents an empirical investigation of unsupervised sentiment classification of Chinese reviews and proposes an algorithm to remove domain-specific sentiment noise words. Section 3.6 introduces the use of substring-group features for sentiment classification through a transductive learning-based algorithm.

Experimental results in multiple languages demonstrate the effectiveness of the algorithm and highlight the superiority of the "tfidf-c" approach for term weighting.

Therefore, this chapter provides a comprehensive overview of various aspects of opinion mining and sentiment analysis, and proposes several innovative methods to address different challenges. From filtering and classifying evaluative opinion sentences to grouping product features, and utilizing topic modeling and similarity measurement for sentiment mining, this chapter covers a wide range of topics and techniques. The practicality and superiority of the proposed methods are demonstrated through empirical experiments. These studies lay a foundation for further advancements in opinion mining and sentiment analysis, and hold significant value in practical applications.

3.1   Identifying Evaluative Sentences in Online Discussions

3.1.1   Introduction

To address the interference of irrelevant information such as emotional statements and personal attacks in online discussions and comments on opinion mining, it is necessary to filter out evaluative opinion sentences that truly express opinions on the topic and its different aspects. Therefore, this section proposes a new unsupervised method to solve this problem. This method uses natural language processing techniques and machine learning algorithms to automatically filter and classify sentences, identify sentences with evaluative opinions, and distinguish them from irrelevant information. Future research can explore more precise and efficient methods for identifying evaluative opinions, and combine the task with sentiment polarity analysis to further improve the effectiveness and application value of opinion mining.

The aim of this work is to identify sentences with evaluative opinions, also known as opinion sentences. Although this problem may seem similar to subjectivity classification, it is actually a completely new problem. Therefore, it has unique challenges and application value in the field of natural language processing. This section focuses on the identification of evaluative sentences and provides a foundation for further sentiment analysis. There are existing studies for this purpose [1–4].

3.1.2   The Proposed Technique

Figure 3.1 provides an overview of the overall process of the proposed technique, which consists of four steps.

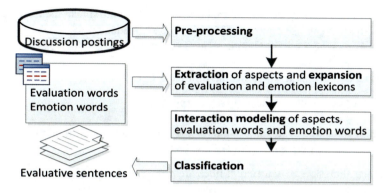

Fig. 3.1 Overview of the proposed technique

Extraction of Aspects and Expansion of Evaluation and Emotion Lexicons

This section introduces a technique for discovering aspects and expanding the given evaluation word list and emotion word list. In this work, we adopt a method similar to the double propagation (DP) method in [5]. DP is a bootstrapping technique that iteratively extracts aspects and expands the initial seed opinion words using some dependency relationships between opinion words and aspects. The input is only a list of opinion words. However, since we focus on evaluation words and emotion words separately, we need to modify the DP method. Specifically, we extend the DP method into two separate processes: emotion word propagation and aspect word propagation. Through these two processes, we can discover and expand evaluation words and aspect words respectively and generate new lists of emotion and aspect words for subsequent sentiment analysis tasks.

DP method uses such relationships to expand the initial seed opinion words and aspect lists iteratively. It uses dependency relationships to extract other potential opinion words and aspects from the known opinion words and aspects. Then, by merging the newly extracted opinion words and aspects with the original opinion words and aspect lists, the next iteration is performed until no new opinion words and aspects are extracted. In this way, we can obtain a more comprehensive aspect vocabulary and a more comprehensive sentiment vocabulary for subsequent sentiment analysis tasks.

In order to build an accurate dependency tree, a reliable Chinese dependency parser is needed. Currently, there are many Chinese dependency parsers to choose from, including both open-source and commercial software. Among them, ICTParser1 and LTP are two common Chinese dependency parsers that provide web demos and have been widely used in practical applications. However, they do not provide a complete API, making it inconvenient for others to use. Additionally, Stanford Parser is also a popular Chinese dependency parser, but according to some experimental results, its performance may not be ideal. Therefore, in this work, researchers adopted a method based on POS tagging to approximate the relations in [5]. This method can use common Chinese word segmentation tools, such as jieba and HanLP, to generate POS tagging sequences for the text, and then use these tags to approximate dependency relationships.

Our adapted technique performs the following tasks:

1. Extract aspects using evaluation or emotion words.
2. Extract aspects using extracted aspects.
3. Extract evaluation words and emotion words using the given or extracted evaluation words and emotion words respectively.

For each subtask above, different rules are proposed:

1. **Rule for Task 1(E → A)**: In this technique, when an evaluation or emotion word E appears near a noun term N, the above tasks are performed to extract the noun term as aspect A. If there is an adjective or another noun term between N and E, then the noun term will not be extracted as an aspect. When two or more noun terms appear near E, the nearest noun term to E will be selected as the extracted aspect. For example, if the evaluation word "Elegant" appears near the noun phrase "red dress" and there is no other adjective or noun phrase between them, then "red dress" will be extracted as aspect A.

   This rule is very intuitive and mainly applicable to evaluation words, as evaluations are typically expressed with respect to a specific aspect. For example, in the sentence (a) below, "weak" is the given evaluation word, and "Argentina" and "defense" are both noun terms. Finally, "defense" is identified as the aspect because it is closer to the given evaluation word "weak" than "Argentina." This rule is also useful for extracting emotion words since emotions are often associated with specific aspects.

   (a) 阿根廷/n 的/u 后防/n 很/d 差/adj (Argentina defense is very weak.)

2. **Rules for Task 2 (A → A)**: There are two rules here:

   If one of the conjoined noun terms is an extracted aspect, then the other noun term is also an aspect. In sentence (b), "勒大(Löw)" and "小伙子们(the players)" are conjoined by the conjunction word "和 (and)." Then, if ""小伙子们 (the players)"has been extracted as an aspect," 勒夫(Löw)" will be extracted as an aspect as well, and vice versa.

   (b) 勒夫 /n 和 /c 小伙子们 J/n 都 /d 很 /d 努力/adj (Löw and the players are both hard-working.)

   If a noun term N appears before or after an extracted aspect A and they are separated by "的," then N is extracted as an aspect. Applying this rule to sentence (a), if "后防 (defense)" is an extracted aspect, then "阿根廷 (Argentina)" is inferred as an aspect A.

3. **Rules for task 3(E → E)**: Again, we have two rules:

   If an adjective term Adj appears within a text window of three words before or three words after a given or extracted evaluation word E, then Adj is extracted as a new evaluation word. Take sentence (c) as an example, if "强(strong)" is a given evaluation word, then the adjective term "主动(proactive)" is extracted as a new evaluation word. "主动(proactive)" is classified as an evaluation word since "强 (strong)" is an evaluation word.

(c) 德国队/n 后防/n 很/d 主动/adj 很/d 强/adj (The German defense is proactive and strong.)

    If an adjective term "Adj" appears within a text window, and the distance between this window and the given or extracted emotion word "E" is no more than three words, we extract the adjective "Adj" as a new emotion word. By employing these separate rules, we can determine whether an evaluation or emotion word has been extracted.

Algorithm 3.1: Algorithm for Discovering Aspects and Expanding Evaluation and Emotion Words Lists

Input: Text corpus: R
        Evaluation word seeds: vas // the given evaluation word lexicon
        Emotion word seeds: mos // the given emotion word lexicon
Output: All evaluation words: VA
        All emotion words: MO
        All aspects: A
1: $VA = vas$; $MO = mos$; $A = \varnothing$
2: $seedVA = vas$; $seedMO = mos$; $seed A = \varnothing$
3: **while** ($seedVA \; != \; \varnothing \;|\; seed MO \; ! \; = \; \varnothing \;|\; seed 1 \; ! \; = \; \varnothing$):
4:      $deltaVA = \varnothing$; $deltaMO = \varnothing$; $deltaA = \varnothing$
5:      **for** each POS-tagged sentence **in** R:
6::   // Task 1
7::   Extract aspects $newA$ using $\mathbf{E} \to \mathbf{A}$ based on $seedVA \cup seedMO$
8::   **Add** the elements in new A but not in A **into** $deltaA$
9:        // Task 2
10:       Extract aspects $newA$ using $\mathbf{A} \to \mathbf{A}$ based on $seedA$
11:       **Add** the elements in $newA$ but not in A **into** $deltaA$
12:       // Task 3
13:       Extract evaluation words $newVA$ using $\mathbf{E} \to \mathbf{E}$ based on $seedVA$
14:       **Add** the elements in $newVA$ but not in $VA \cup MO$ **into** $deltaVA$
15:       Extract emotion words $newMO$ using $\mathbf{E} \to \mathbf{E}$ based on $seedMO$
16:       **Add** the elements in $newMO$ but not in 2: $VA \cup MO$ **into** $deltaMO$
17:   **Add** $deltaVA$ **into** VA
18:   **Add** $deltaMO$ **into** MO
19:   **Add** $deltaA$ **into** A
20:   $SeedVA = deltaVA$; $seed MO = deltaMO$; $seed A = deltaA$

Based on the above description, the detailed algorithm is given in Algorithm 3.1, which is self-explanatory.

Aspects, Evaluation Words, and Emotion Words Interaction

We believe that in evaluative sentences, emotion words and evaluation words are more likely to appear, while in nonevaluative sentences, their occurrence is lower. Therefore, we can use this distributional feature to increase the weight of these words in evaluative sentences. At the same time, we can also consider that a word may have different usage in different domains, so we need to reclassify evaluation words and emotion words according to different application domains to reduce the impact of errors.

To better explain the IAEE (Interaction of Aspect, Evaluation, and Emotion) formula, we use a directed tripartite graph in Fig. 3.2 to model the relationships between the three concepts. Each concept is represented as a node, and their relationships are represented as edges. We can see that the aspect node is connected to both the evaluation word and emotion word nodes, and there is also a connection between the evaluation word node and emotion word node. These edges represent the interactions between different concepts. In the IAEE formula, we consider these interactions. We treat aspect, evaluation word, and emotion word as three different layers, and introduce the emotion word layer as a new third layer. The role of the emotion word layer is to inhibit the relationship between the evaluation word and aspect, thereby reducing the influence of evaluation words and emotion words on the aspect weights. This is something that is not considered in the HITS algorithm [6]. Therefore, the IAEE formula can more accurately model the relationships between these three concepts, improving the accuracy and precision of evaluating sentence classification.

Formally, the tripartite graph is represented as $G = \langle V_a, V_{va}, V_{mo}, E_{va-a}, E_{mo-a} \rangle$, where $V_a = \{a_i\}$, $V_{va} = \{va_j\}$, $V_{mo} = \{mo_k\}$ are aspects, evaluation words and emotion words, respectively; E_{va-a} denotes the relationship between V_{va} and V_a; E_{va-a} denotes the relationships between V_{mo} and V_a.

Here, the relationship refers to co-occurrence in a sentence. That is, if an aspect $a \in V_a$ and an evaluation (or emotion) word $v \in V_{va}$ (or $m \in V_{mo}$) co-occurs in a sentence, a directed edge (v, a) (or (m, a)) is created. The edges are all of unit weight, that is, multiple occurrences are considered as 1.

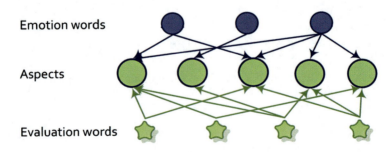

Fig. 3.2 Interaction modeling of aspects, evaluation words, and emotion words (IAEE)

The asp(a_i) can be calculated using the Eq. (3.1), which defines the method for calculating the aspect score. The aspect score is directly proportional to the sum of the evaluation word scores and inversely proportional to the sum of the emotion word scores. Specifically, assuming that a sentence contains n evaluation words, the evaluation score of the ith valuation word is eva(va$_j$), and m emotion words, the emotion score of the jth emotion word is emo(m_k), then the aspect score asp(a_i) can be calculated as follows:

$$\text{asp}(a_i) = \lambda \times \sum_{(i,j)\in E_{va-a}} \text{eva}(va_j) - (1-\lambda) \times \sum_{(i,k)\in E_{mo-a}} \text{emo}(m_k) \tag{3.1}$$

In the formula, λ is the damping factor, which is used to adjust the relative influences of evaluation words and emotion words on the aspect score. When λ is 0, only the contribution of emotion words to the aspect score is considered; when λ is 1, only the contribution of evaluation words to the aspect score is considered. In this study, λ is set to the default value of 0.5, indicating that the influence of evaluation words and emotion words on the aspect score is equal.

Since aspects and evaluation words mutually reinforce each other, the score eva of an evaluation word is computed with Eq. (3.2), where a_i is an associated aspect with the evaluation word va$_j$.

$$\text{eva}(va_j) = \sum_{(i,j)\in E_{va-a}} \text{asp}(a_i) \tag{3.2}$$

The computation of score emo(mo_k) for each emotion word mo_k involves a certain process. To consider the inhibiting effect, we first introduce an intermediate score tmp(mo_k), which is defined by Eq. (3.3). Since the emo score indicates nonevaluative strength, the score for an emotion word emo(mo_k) should have opposite effect of tmp(mo_k). That is, the larger the tmp(mo_k) is, the smaller the emo(mo_k) should be as shown in Eq. (3.4). To achieve the desired effect of emo (mo_k), we define it with Eq. (3.5), where max represents the maximum value of tmp (m_k) of all emotion words (see Eq. (3.6)).

By introducing the intermediate score tmp(mo_k), we are able to consider the mutual influence between emotion words. A larger value of emo(mo_k) indicates a stronger inhibiting effect of that emotion word, resulting in a corresponding decrease in the score of emo(mo_k). This design allows us to more accurately reflect the relative strength and correlation between emotion words.

$$\text{tmp}(mo_k) = \sum_{(i,k)\in E_{mo-a}} \text{asp}(a_i) \tag{3.3}$$

$$\mathrm{emo}(\mathrm{mo}_k) \propto -\mathrm{tmp}(\mathrm{mo}_k) \tag{3.4}$$

$$\mathrm{emo}(\mathrm{mo}_k) = -\mathrm{tmp}(\mathrm{mo}_k) + m = m - \mathrm{tmp}(\mathrm{mo}_k) \tag{3.5}$$

$$m = m\{\mathrm{tmp}(\mathrm{mo}_1), \mathrm{tmp}(\mathrm{mo}_2), \ldots, \mathrm{tmp}(\mathrm{mo}_{|V_{\mathrm{mo}}|})\} \tag{3.6}$$

To further understand Eqs. (3.3), (3.4), (3.5), and (3.6), let us discuss two extreme cases. If $\mathrm{tmp}(\mathrm{mo}_k)$ is very high, which means that the aspects are strong because it is computed from aspects in Eq. (3.3), then the emotion score should be low. This is reflected by Eq. (3.5). The strong aspects are caused by strong connections with evaluation words due to their positive mutual reinforcements. If $\mathrm{tmp}(\mathrm{mo}_k)$ is very low, which means the aspect scores are low because it is computed in Eq. (3.2), At this point, the emotion score should be high. This is also reflected in Eq. (3.5). When the emotion words are strong, the associated aspects will be suppressed, resulting in lower aspect scores, and vice versa. Eq. (3.1) precisely achieves the adjustment of the emotion score in this manner.

Algorithm 3.2: Iterative Computation of the IAEE Model

Input: Evaluation words: *VA*
　　　　Emotion words: *MO*
　　　　Aspects: *A*
　　　　Co-occurrence relationship between *VA* and *A* : $E_{\mathrm{va} - \mathrm{a}}$
　　　　Co-occurrence relationship between *MO* and *A* : $E_{\mathrm{mo} - \mathrm{a}}$
Output: Evaluative scores of *VA*, *MO* and *A*
1: Initialize the scores of asp, eva and emo to 1;
2: **Repeat** 50 times:
3:　　update each asp(a_i) using Eq. (3.1)
4:　　update each eva (va_j) using Eq. (3.2)
5:　　update each emo (mo_k) using Eq. (3.5)
6:　　normalize asp(a_1), asp(a_2), ..., asp(a_I) to [0, 1]
7:　　normalize eva(va_1), eva(va_2), ..., eva(va_J) to [0, 1]
8:　　normalize emo(mo_1), emo(mo_2), ..., emo(mo_K) to [0, 1]

To solve these equations, we employ the classic power iteration method. Power iteration is a technique used to calculate the largest eigenvalue and its corresponding eigenvector of a matrix. It has found wide applications in natural language processing and sentiment analysis. Detailed algorithm steps are provided in Algorithm 3.2. When using this method, we need to prepare some input data. First, we have the text corpus, which contains the text samples we want to analyze. Next, we have the evaluation words VA, which are vocabulary used for sentiment evaluation of the text. The emotion words MO are vocabulary that describes emotional states and help us capture the sentiment nuances in the text. Lastly, we define the aspects A, which represent specific interests or topics we are interested in, such as product performance or service quality. The algorithm outputs individual scores for each aspect, evaluation word, and emotion word. By iteratively computing the input data, we can obtain importance scores for each aspect in the text, as well as weights for each evaluation word and emotion word in different aspects. These scores and

weights can assist us in understanding the sentiment orientation and aspect preferences in the text. In our experiments, we have chosen to run the algorithm for
50 iterations. Through multiple iterations, we can progressively optimize the model's
performance and ensure the stability and accuracy of the results. Based on experimental validation, we believe that conducting 50 iterations is sufficient to obtain
reliable results.

Classification

Given the scored evaluation words, emotion words, and aspects, and the corpus, this
step classifies each sentence.

1. **Task 1**: It matches all aspects $\{a_1, \ldots, a_l\}$ in the sentence s and finds the highest
   evaluative score top A of an aspect in s (Eq. (3.7)). If top A is greater than a
   predefined threshold ρ (the default is 0.6), we proceed to step 2; Otherwise, the
   sentence s is classified as nonevaluative.

$$\text{top } A = \max_{1 \leq i \leq l} \text{asp}(a_i) \tag{3.7}$$

   For example, for the sentence "German defense is proactive and strong," this
   step first finds the only aspect "German defense." Assume its asp score's higher
   than 0.6, we go to step 2.

2. **Task 2**: It matches all evaluation words $\{va_1, \ldots, va_J\}$ and emotion words
   $\{mo_1, \ldots, mo_K\}$ in the sentence, and then sums up the evaluation word scores
   vaSum (Eq. (3.8)) and emotion word scores moSum (Eq. (3.9)). If vaSum is
   greater than moSum, sentence s is classified as evaluative; Otherwise,
   nonevaluative.

$$\text{vaSum} = \sum_{1 \leq j \leq j} \text{eva}(va_j) \tag{3.8}$$

$$\text{moSum} = \sum_{1 \leq k \leq K} \text{emo}(mo_k) \tag{3.9}$$

   Following the above example, we have two evaluation words "proactive" and
"strong," which result in vaSum > 0; there is no emotion word, resulting in
moSum $= 0$. This sentence is thus classified as an evaluative sentence.

3.1.3 Experiments

We used four datasets to evaluate the proposed IAEE system. These datasets cover the following topics: (1) 2010 FIFA, (2) 2010 NBA, (3) Guo Degang's dispute with Beijing TV, and (4) Tang Jun's fake PhD degree. These datasets contain a large amount of discussions related to these topics. To evaluate the datasets, we employed two PhD students in computer science to annotate each sentence as evaluative or nonevaluative. They made judgments based on the content and expression of the sentences to determine if they had evaluative nature. This annotation process aimed to provide reliable labeled data for evaluating the performance of the system. To assess the inter-rater agreement among the evaluators, we calculated the Kappa scores. Kappa score is a statistical measure used to quantify the level of agreement among evaluators, with a value closer to 1 indicating higher agreement. In our experiments, the Kappa scores among the evaluators ranged from 0.841 to 0.894, indicating almost perfect agreement. To gain a more detailed understanding of these datasets, we provided the detailed information in Table 3.1. The table includes the names of the datasets, the topics they cover, the amount of data, and other relevant information.

Methods and Settings

The proposed IAEE algorithm is compared with six baseline methods, which are categorized into supervised and unsupervised approaches. We first present the supervised methods.

1. **NB**: This method employs a Naive Bayesian classifier to perform classification. It assumes the independence of features and assigns evaluative or nonevaluative labels to sentences based on their probabilistic likelihood.
2. **SVM**: This method utilizes Support Vector Machines for classification. SVM constructs a hyperplane to separate evaluative and nonevaluative sentences, demonstrating strong classification performance.

    For NB and SVM, we use Chinese segmented words as features for training and testing. All results are obtained through tenfold cross-validation, ensuring the evaluation's accuracy and reliability.

    Next, we describe several supervised methods based on rules and lexicons:

Table 3.1 Summary of the four datasets

	#Postings	#Sentences		Kappa
		Evaluative	Nonevaluative	
ARG vs. GER(FIFA)	1672	1393	1607	0.894
Lakers vs. Celtics(NBA)	1984	883	2117	0.881
Guo Degang(GD)	2196	682	2318	0.847
Tang Jun (TJ)	1712	1115	1885	0.841

3. **Lexi**: This method uses the original evaluation and emotion lexicons from HowNet. Based on the relative count of evaluation words and emotion words in a sentence, it classifies the sentence as evaluative or nonevaluative.
4. **E-Lexi**: Similar to Lexi, this method considers expanded evaluation and emotion words. By broadening the coverage of vocabulary, it enhances the ability to capture the sentiment nature of sentences.
5. **A-E-Lexi**: Building upon E-Lexi, this method incorporates extracted aspects. If a sentence contains at least one aspect and has more evaluation words than emotion words, it is classified as evaluative; otherwise, it is classified as nonevaluative. If a sentence lacks aspects or evaluation words, it is classified as nonevaluative.
6. **Double-HITS**: This method leverages extracted aspects, given and expanded evaluation words, and emotion words. Two HITS algorithms are separately applied: evaHITS operates on aspects (authorities) and evaluation words (hubs), while emoHITS operates on aspects (authorities) and emotion words (hubs). Scores such as evaAuth and emoAuth for each aspect, evaHub for each evaluation word, and emoHub for each emotion word are obtained. During classification, a sentence is classified as evaluative if it satisfies contains aspects $\{a_1, \ldots, a_I\}$, evaluation words $\{va_1, \ldots, va_J\}$ and emotion words $\{mo_1, \ldots, mo_K\}$, otherwise, it is classified as nonevaluative.

$$\max_{1 \leq i \leq I} \{\text{evaAuth}(a_i)\} > \max_{1 \leq i \leq I} \{\text{emoAuth}(a_i)\} \tag{3.10}$$

$$\sum_{1 \leq j \leq J} \{\text{evaHub}(va_j)\} > \sum_{1 \leq k \leq K} \{\text{evaHub}(mo_k)\} \tag{3.11}$$

Evaluation Results

The comparison results are presented in Table 3.2, where Avg represents the average performance across the 4 datasets. Below, we provide a detailed discussion of the observations:

- The proposed IAEE method achieves the highest F1-score overall, surpassing all other methods significantly.
- The fully supervised methods NB and SVM demonstrate poor performance in terms of F1-score. We believe this is mainly because the key determining factors for evaluative sentences are aspects and evaluation words, which are challenging to detect using these two supervised techniques.
- In terms of F1-score, Lexi performs worse than E-Lexi, and E-Lexi performs worse than A-E-Lexi. The reason is that Lexi only utilizes original words from HowNet for classification and does not incorporate any expanded evaluation words or emotion words. E-Lexi outperforms Lexi by utilizing the expanded

Table 3.2 Comparison results

	F1-score					Precision					Recll				
	FIFA	NBA	GD	TJ	Avg	FIFA	NBA	GD	TJ	Avg	FIFA	NBA	GD	TJ	Avg
NB	0.76	0.63	0.43	0.61	0.61	0.75	0.60	0.48	0.61	0.61	0.76	0.66	0.40	0.62	0.61
SVM	0.74	0.53	0.30	0.57	0.53	0.81	0.77	0.68	0.74	0.75	0.68	0.40	0.20	0.46	0.44
Lexi	0.68	0.53	0.43	0.59	0.56	0.61	0.39	0.29	0.46	0.44	0.77	0.84	0.79	0.85	0.81
E-Lexi	0.73	0.66	0.52	0.69	0.65	0.71	0.52	0.38	0.56	0.54	0.74	0.91	0.85	0.91	0.85
A-E-Lexi	0.74	0.70	0.53	0.76	0.68	0.77	0.58	0.39	0.66	0.60	0.71	0.89	0.84	0.89	0.83
Double-HITS	0.75	0.72	0.59	0.76	0.70	0.81	0.64	0.48	0.68	0.65	0.70	0.83	0.76	0.86	0.79
IAEE	0.75	0.81	0.69	0.78	0.76	0.83	0.81	0.64	0.70	0.75	0.67	0.81	0.74	0.88	0.78

evaluation and emotion words. A-E-Lexi, which incorporates aspect information, performs even better.

- Double-HITS outperforms all methods above it in terms of F1-score. We believe this is because Double-HITS is capable of re-weighting aspects, evaluation words, and emotion words, thereby partially addressing the interaction among these three concepts.
- The IAEE method and A-E-Lexi are similar, but IAEE incorporates weighted aspects, evaluation words, and emotion words. IAEE improves the F1-score of A-E-Lexi by an average of 8%. We observe that the precision of A-E-Lexi is much lower than that of IAEE, but the recall is better.
- IAEE also outperforms Double-HITS in terms of F1-score. Both methods employ the same final classification strategies. The reason for IAEE's superior performance is its ability to fully consider the interaction among the three concepts within a single framework, whereas Double-HITS treats them separately and fails to leverage the interaction between evaluation and emotion through aspects.

In summary, the proposed IAEE method outperforms all the baseline methods in terms of evaluation metrics. Firstly, IAEE demonstrates excellent performance in terms of F1-score, significantly surpassing other methods. Its efficient feature extraction and accurate classification ability enable it to better identify evaluative sentences. Secondly, compared to the fully supervised NB and SVM methods, IAEE exhibits stronger capabilities in recognizing evaluation words and aspects, which are considered key factors in determining evaluative sentences. In comparison to Lexi, E-Lexi, and A-E-Lexi, IAEE not only incorporates expanded evaluation words and emotion words but also considers weighted aspect information, leading to improved performance in the classification of evaluative sentences. Additionally, compared to Double-HITS, IAEE comprehensively takes into account the interactions among aspects, evaluation words, and emotion words, achieving more accurate classification through a unified framework. Overall, IAEE's performance in various aspects sets it apart in the field of evaluative sentence classification, demonstrating significant advantages in the conducted experiments.

Influence of the Parameters

The proposed IAEE method has two parameters: the damping factor λ and the evaluative score threshold ρ. We will now demonstrate the influences of their values on the overall performance. In Fig. 3.3, we observe that IAEE achieves the best results (averages over the 4 datasets) when λ is around 0.5, indicating that evaluative and emotion words should carry equal weight in the model. We conducted experiments using a range of ρ values and found similar trends for λ. For the purpose of illustration, we chose $\rho = 0.6$ in Fig. 3.3. Furthermore, in Fig. 3.4, we further investigate the case when ρ is set to 0.6 (with $\lambda = 0.5$), where IAEE achieves the highest F1-score. The results from Figs. 3.3 and 3.4 reveal the significant impact of parameter selection on the performance of the IAEE method. In our experiments,

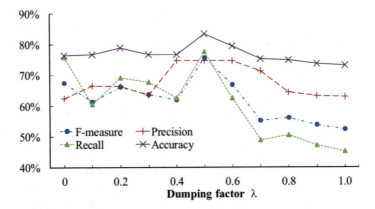

Fig. 3.3 Influence of λ on IAEE

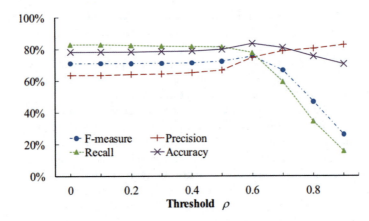

Fig. 3.4 Influence of ρ on IAEE

IAEE demonstrates the best overall performance when λ is close to 0.5, suggesting the importance of balancing the significance of evaluative and emotion words within the model. Additionally, when ρ is set to 0.6, IAEE achieves the optimal F1-score, indicating that 0.6 is a suitable choice for setting the evaluative score threshold.

These findings highlight the sensitivity of the IAEE method to parameter settings. To achieve optimal performance, we recommend conducting parameter tuning in specific applications and considering the characteristics of the task at hand to determine appropriate parameter values. By fine-tuning the parameters with precision, the accuracy and effectiveness of the IAEE method can be further improved, enabling better adaptation to various evaluation tasks and datasets.

3.1.4   Conclusion

This section introduces the problem of identifying evaluative sentences from online discussions. To our knowledge, this problem has not been extensively studied, yet it holds great importance for practical applications. We propose a novel unsupervised method to address this problem, which eliminates the need for laborious manual labeling of training data required in supervised learning. Extensive experiments conducted on real-life discussions demonstrate the effectiveness of our proposed method, surpassing the performance of supervised baselines. In various domains such as social media analysis, product reviews, and sentiment analysis, accurately identifying evaluative sentences enables us to understand user sentiment, assess the quality of products or services, and monitor public opinion trends.

Compared to traditional supervised learning approaches, our unsupervised method offers distinct advantages. By leveraging textual features and structural information present in online discussion data, we employ automated algorithms to infer evaluative sentences. This unsupervised approach provides broader applicability and can be employed across different domains and contexts. The extensive experiments validate the superior performance of our proposed method on real-life online discussion datasets. In comparison to supervised baselines, our method not only achieves comparable or even superior performance but also saves significant human and time resources. The experimental results highlight the accuracy and efficiency of our approach.

Furthermore, we investigate key parameters that influence the method's performance, such as threshold and damping factor. By adjusting these parameters, users gain the flexibility to fine-tune the method according to their specific needs, striking a balance between precision and recall. This customization empowers our method with greater adaptability and caterability to different application scenarios.

In conclusion, the proposed unsupervised method in this section offers an innovative solution for identifying evaluative sentences in online discussions. By eliminating the reliance on manual labeling of training data, our method holds practical value and exhibits excellent performance and flexibility in real-world applications. It provides valuable insights and references for further research and application in this field.

3.2   Grouping Product Features Using Semisupervised Learning with Soft-Constraints

3.2.1   Introduction

In opinion mining of product reviews, it is often desired to generate an opinion summary based on product features/attributes. However, for the same feature, people may express it using different words and phrases. To create a meaningful summary, these words and phrases, which are domain synonyms, need to be grouped together under the same feature group.

This section presents a novel subproblem in opinion mining, namely, the grouping of feature expressions within the context of semisupervised learning. While existing methods exist for solving this problem using unsupervised learning, we argue that some form of user supervision is necessary in practical applications to guide the system in understanding user preferences. We employ an EM formulation to address the problem and enhance it with two soft constraints. These constraints help guide the EM algorithm to generate better solutions. We acknowledge that these constraints can be relaxed during the process to correct any imperfections.

Experimental results using reviews from five different domains demonstrate the competence of the proposed method for this task. The method effectively groups feature expressions and generates accurate opinion summaries. Furthermore, we observe that the method exhibits a degree of generality and robustness when applied to different domains and languages.

3.2.2 The Proposed Algorithm

In our research context, our problem can be viewed as a form of semisupervised learning. We use certain terms to describe the setting of the problem. Firstly, we have a set C of classes, which represent the feature groups under which we want to group the feature expressions. Next, we denote the labeled examples (i.e., the feature expressions with known class labels or seeds) as L, and the unlabeled examples (i.e., the feature expressions without labels) as U. Our goal is to build a classifier using L and U to classify each example in U into its corresponding class.

To address this problem, we can apply various existing algorithms. In this study, we choose to use the EM algorithm due to its efficiency and its ability to easily incorporate prior knowledge. The EM algorithm is an iterative algorithm that estimates the parameters of the model by alternating between the "Expectation" (E-step) and "Maximization" (M-step) steps. In the E-step, it calculates the probabilities of each unlabeled example belonging to each class based on the current parameter estimates, and in the M-step, it updates the model parameters using these probabilities. Through repeated iterations of the E-step and M-step, the EM algorithm gradually optimizes the model parameters, enabling the model to better classify the unlabeled examples.

However, in our research, we propose an augmented EM algorithm to further improve the performance of the classifier. We introduce two soft constraints that guide the execution of the EM algorithm and help generate better solutions. The purpose of these constraints is to incorporate some prior knowledge or restrictions into the process of grouping the feature expressions, thereby enhancing the accuracy and stability of the model.

Semisupervised Learning Using EM

EM algorithm is a popular iterative algorithm for maximum likelihood estimation in problems with missing data. In our research context, our problem involves grouping feature expressions, where the group memberships of unlabeled expressions are considered missing as they lack corresponding group labels. To address this problem, we utilize the EM algorithm combined with the principles of naive Bayesian classification.

In the EM algorithm, we first learn a classifier, denoted as f, using the labeled data. This classifier establishes associations between feature expressions and their respective groups. Subsequently, we apply this classifier to assign probabilistic labels to the unlabeled expressions, indicating the likelihood of belonging to each group. Next, we relearn the classifier f using both the labeled data and the unlabeled data with probabilistic labels. This step aims to obtain more accurate estimates of the feature distributions and conditional probabilities for each group. The iterative process continues until the classifier's performance converges.

In the equations, we employ specific symbols to represent relevant concepts. Given a set of training documents D, each document d_i is treated as an ordered list of words. The symbol $w_{d_i,k}$ represents the kth word in document d_i, with each word drawn from the vocabulary $V = \{w_1, w_2, \ldots w_{|\zeta|}\}$. Our objective is to group feature expressions into predefined groups represented by the set $C = \{c_1, c_2, \ldots, c_{|C|}\}$. The symbol N_{ti} denotes the frequency of word w_t occurring in document d_i, which is used to calculate feature frequencies and conditional probabilities.

$$P(w_t \mid c_j) = \frac{1 + \sum\limits_{i=1}^{|D|} N_{ti} P(c_j \mid d_i)}{|V| + \sum\limits_{m=1}^{|V|} \sum\limits_{i=1}^{|D|} N_{mi} P(c_j \mid d_i)} \qquad (3.12)$$

$$P(c_j) = \frac{1 + \sum\limits_{i=1}^{|D|} P(c_j \mid d_i)}{|C| + |D|} \qquad (3.13)$$

$$P(c_j \mid d_i) = \frac{P(c_j) \prod\limits_{k=1}^{|d_i|} P(w_{d_i,k} \mid c_j)}{\sum\limits_{r=1}^{|C|} P(c_r) \prod\limits_{k=1}^{|d_i|} P(w_{d_i,k} \mid c_r)} \qquad (3.14)$$

For our problem, the surrounding words contexts of the labeled seeds form L, while the surrounding words of the nonseed feature expressions form U. When EM converges, the classification labels of the unlabeled feature expressions give us the final grouping.

In our problem, we aim to group feature expressions that have similar semantic or feature characteristics. To achieve this, we utilize a set of labeled seed expressions where the surrounding word contexts are used to construct the labeled data set L. Additionally, we have a set of unlabeled feature expressions where the surrounding word contexts form the unlabeled data set U.

By applying the EM algorithm, we iteratively adjust the classification labels of the unlabeled feature expressions to assign them to the same groups as the labeled seed expressions that exhibit similar semantic or feature properties. The EM algorithm updates the model parameters and reassigns the classification labels based on the current classification results and the known seed information until convergence.

When the EM algorithm converges, the final classification labels of the unlabeled feature expressions provide us with the definitive grouping results. This means that through the iterative optimization process of the EM algorithm, we are able to automatically assign the unlabeled feature expressions to the groups that are similar to the labeled seed expressions, thereby accomplishing the task of feature expression grouping.

Proposed Soft-Constrained EM

Although EM can be directly applied to deal with our problem, we can do better. As we discussed earlier, EM only achieves a local optimum based on the initialization, that is, the labeled examples or seeds. We propose a method to improve the initialization by utilizing natural language constraints. We introduce additional seeds called soft-labeled examples or soft seeds (SL) that are likely to be correct. Soft-labeled examples are treated differently from the original labeled examples in L. With the incorporation of soft seeds, we present a modified version of EM called Soft-Constrained EM (SC-EM). By introducing soft constraints, we can better guide the iterative process of EM, thereby enhancing the algorithm's performance and accuracy.

Algorithm 3.3: The Proposed SC-EM Algorithm

Input:
– Labeled examples L
– Unlabeled examples U
1: Extract SL from U using constraints;
2: Learn an initial naïve Bayesian classifier f_0 using $L \cup SL$ and Eqs. (3.12) and (3.13);
3: **repeat**
4:      // E-Step
5:      **for** each example d_i in U (including SL) do
6:          Using the current classifier f_x to compute $P(c_j \mid d_i)$ using Eq. (3.14)
7:      **end**
8:      // M-Step
9:      Learn a new naïve Bayesian classifier f_x from L and U by computing
         $P(w_t \mid c_j)$ and $P(c_j)$ using Eqs. (3.12) and (3.13).
10: **until** the classifier parameters stabilize
Output: the classifier f_x from the last iteration

Compared to the original EM algorithm, the SC-EM algorithm has two main differences:

- In SC-EM, we introduce soft constraints to improve the initialization of the algorithm. Using these soft constraints, we identify potential correct seed samples within the unlabeled examples, forming a set of soft-labeled examples called SL. By increasing the size of the training set, we can achieve better results. These soft-labeled examples are treated similarly to the labeled examples during the initial stage of the algorithm.
- In the subsequent iterations of SC-EM, the soft-labeled examples SL and the unlabeled examples U are considered together as training data. This means that the classifier predicts the labels of the unlabeled examples in each iteration. Afterwards, a new classifier is built using both L and U_{PL}, which is U with probabilistic labels. Unlike the labeled examples, the class labels of the soft-labeled examples can change because the constraints may introduce errors. Through this iterative process, the SC-EM algorithm gradually optimizes the model parameters to achieve better classification results. The detailed algorithmic details are provided in Algorithm 3.3.

3.2.3 Generating SL Using Constraints

As mentioned earlier, two forms of constraints are used to induce the soft-labeled set SL. For easy reference, we reproduce them:

1. Feature expressions sharing some common words are likely to belong to the same group. For example, if multiple feature expressions have similar words appearing in the text, we can infer that they may belong to the same category or group.
2. Feature expressions that are defined as synonyms in a dictionary are likely to belong to one group. If multiple feature expressions have the same meaning or synonyms, we can infer that they may belong to the same category or group

The purpose of these constraints is to generate soft-labeled examples that are likely to belong to the correct groups among the unlabeled feature expressions. By combining these constraints with the iterative process of the EM algorithm, we can gradually optimize the classifier to obtain more accurate classification results.

Algorithm 3.4: Generating the Soft-labeled Set SL

1: **for** each feature expression $u \in U$ **do**
2:       **for** each feature expression $L_i \in L$ **do**
3:          score(L_i)← 0;
4:          **for** each feature expression $e \in L_i$ **do**
5:             **if** u is a single word expression **then**
6:                **if** e is a single word expression **then**
7:                   **if** u and e are synonyms **then**
8:                      score(L_i)← score(L_i) + 1;

(continued)

9:	**else if** $w \in e$ **then** // e is a phrase
10:	$score(L_i) \leftarrow score(L_i) + 1$
11:	else // u is a phrase
12:	**if** e is a single word expression **then**
13:	**if** $e \in u$ **then** // u is a phrase
14:	$score(L_i) \leftarrow score(L_i) + 1$
15:	**else**
16:	$s \leftarrow e \cap u$;
17:	$score(L_i) \leftarrow score(L_i) + \mid s \mid$
18:	u is added to SL_j s. t. $argmax_{L_j} score(L_j)$

According to the number of words in the feature expressions, they can be divided into word expressions and phrase expressions, which need to be processed differently. Here is a detailed algorithm (Algorithm 3.4) for handling these expressions. In the algorithm, the labeled set L consists of multiple subsets (e.g. L_1, $L_2 L_3$, etc.), with each subset corresponding to a specific category or feature group. Each subset contains a set of labeled examples (i.e., feature expressions). Similarly, the output set SL (i.e., soft label set) also consists of multiple subsets, with each subset corresponding to a category or feature group and containing a set of examples with soft labels.

The goal of the algorithm is to soft label the unlabeled examples in the unlabeled set U by comparing each feature expression in the unlabeled set with the feature expressions in the labeled subsets L_i, based on specific constraint conditions. The constraint conditions include cases where feature expressions share the same words and synonyms in the dictionary. If an unlabeled example satisfies any constraint condition, it implies that it likely belongs to the corresponding category or feature group, and it is added to the respective soft label set SL_i.

However, sometimes an unlabeled example may satisfy constraint conditions from multiple labeled subsets simultaneously, which requires resolving conflicts. For example, a word may be a synonym in multiple feature groups, and in such cases, we need to determine which feature group it is more likely to belong to. Similarly, an unlabeled example may be a synonym for multiple word expressions in a labeled subset. To resolve these conflicts, the algorithm makes judgments based on different situations, such as whether it is a word or phrase and whether there are shared words. Ultimately, by calculating scores, it determines the degree of association between the unlabeled example and a specific category and assigns it to the category with the highest score.

3.2.4   Distributional Context Extraction

The process of preparing a document d_i is required for each feature expression e_i in naive Bayes learning in order to apply the proposed algorithm. The document d_i is generated by aggregating the distributional context of each sentence s_{ij} in the corpus that contains the expression e_i. The context of a sentence refers to the surrounding words of e_i in a text window of $[-t, t]$, where t is a specified window size that includes the words in e_i. Given a relevant corpus R, the algorithm in Algorithm 3.5 is used to generate the corresponding document d_i for each feature expression e_i in the label set L (or U). Stopwords, commonly ignored words in text analysis, are removed during document generation. The aim of this algorithm is to soft-label unlabelled examples by analyzing shared and synonymous relationships between feature expressions, to assist with further classification and analysis work.

Algorithm 3.5: Distributional Context Extraction

1: **for** each feature expression e_i in L (or U) **do**
2:      $S_i \leftarrow$ all sentences containing e_i in R;
3:      **for** each sentence $s_{ij} \in S_i$ **do**
4:          $d_{ij} \leftarrow$ words in a window of $[-t, t]$ on the left and right (including the words in e_i);
5:      $d_i \leftarrow$ words from all $d_{ij}, j = 1, 2, \ldots, |S_i|$;
              // duplicates are kept as it is not union

For example, a feature expression from L (or U) is $e_i =$ "screen" and there are two sentences in our corpus R that contain "screen."
$s_{i1} =$ "The LCD screen gives clear picture."
$s_{i2} =$ "The picture on the screen is blur."
We use the window size of $[-3, 3]$. Sentence s_{i1}, gives us $d_{i1} = <$ LCD, screen, give, clear, picture $>$ as a bag of words. "the" and "is" are removed as stopwords. s_{i2} gives us $d_{i2} = <$ picture, screen, blur$>$. "on," "the" and "is" are removed as stopwords. Finally, we obtain the document d_i for feature expression e_i as a bag of words:
$d_i = <$ LCD, screen, give, clear, picture, picture, screen, blur $>$.

3.2.5 Experiments

This section provides a detailed evaluation of the SC-EM algorithm and presents a comprehensive comparison with other major existing methods for solving the current problem. During the evaluation process, the SC-EM algorithm is thoroughly analyzed and empirically studied, taking into account the specific requirements and performance metrics of the problem.

Review Data Sets and Gold Standards

To validate the wide applicability of the proposed method, we conducted a series of experiments using review datasets from five different domains: Home Theater, Insurance, Mattress, Car, and Vacuum Cleaner. These datasets were provided by a

Table 3.3 Data sets and gold standards

	Home theater	Insurance	Mattress	Car	Vacuum
#Sentences	6355	12,446	12,107	9731	8785
#Reviews	587	2802	933	1486	551
#Feature expressions	237	148	333	317	266
#Feature groups	15	8	15	16	28

company specializing in opinion mining services, ensuring the accuracy and reliability of the data. Please refer to Table 3.3 for detailed information.

Evaluation Measures

The SC-EM algorithm is based on semisupervised learning and can be used to address classification and clustering problems. In order to evaluate the performance of this algorithm, we can also consider SC-EM as a clustering method with initial seeds. Hence, we can utilize clustering evaluation methods to assess the performance of SC-EM in clustering tasks.

Given a dataset DS containing multiple data points, we can divide DS into k disjoint subsets, denoted as $DS_1, \ldots, DS_i, \ldots, DS_k$, based on a known number of clusters, k. This partitioning is based on the gold standard partition $G = \{g_1, \ldots, g_j, \ldots, g_k\}$ of the dataset. Each subset represents a cluster, and the gold standard partition determines the correct cluster membership for each data point.

1. **Entropy**: When evaluating the clustering results, we can use entropy to measure the uncertainty of each cluster. For each cluster DS_i, we can calculate its entropy value using Eq. (3.15). In the equation, $P_i(g_j)$ represents the proportion of data points in DS_i that belong to cluster g_j. By computing the entropy for each cluster, we can obtain information about the richness of the clustering result . In addition to the entropy values for individual clusters, we can also calculate the total entropy of the entire clustering result. The total entropy considers the entropy values of all clusters and is computed using Eq. (3.16). This metric provides an assessment of the information content of the overall clustering result, where a lower total entropy indicates a more distinct and pure clustering result.

$$\text{entropy}\,(DS_i) = -\sum_{j=1}^{k} P_i(g_j) \log_2 P_i(g_j) \tag{3.15}$$

$$\text{entropy}_{\text{total}} = \sum_{i=1}^{k} \frac{|DS_i|}{|DS|} \,\text{entropy}\,(DS_i) \tag{3.16}$$

2. **Purity**: Purity is a metric used to evaluate the quality of clustering results. For each cluster, we can calculate its purity value using Eq. (3.17). Total purity is a

comprehensive metric for evaluating the overall clustering result. It takes into account the purity values of all clusters and is calculated using Eq. (3.18)

$$\text{purity}_{(DS_i)} = \max_{j} P_i(g_j) \tag{3.17}$$

$$\text{purity}_{\text{total}} = \sum_{i=1}^{k} \frac{|DS_i|}{|DS|} \text{ purity } (DS_i) \tag{3.18}$$

In the testing process, the unlabeled set U is also used as our test set. This is reasonable because our goal is to assign the unlabeled data to their respective categories.

Baseline Methods and Settings

The proposed SC-EM method is compared with a set of existing methods, which can be categorized into unsupervised and semisupervised methods. We first list the unsupervised methods, which are particularly useful when dealing with unlabeled data.

1. **LDA**: LDA is a popular topic modeling method widely used in the field of text analysis. Given a set of documents, LDA extracts combinations of words corresponding to different topics. In our case, each feature expression can be viewed as a word, and the documents represent the distributional contexts of each feature expression.
2. **mLSA**: mLSA is an advanced unsupervised method for solving the classification problem of feature expressions. It is based on the LDA method and has been extensively discussed and improved in related research.
3. **K-means**: K-means is a clustering method based on cosine similarity commonly used in unsupervised learning for data clustering tasks. By computing the similarity between data points, K-means divides the data points into different clusters.

In the semisupervised methods, we further classify them into three subclasses: unconstrained, hard-constrained, and soft-constrained methods. The unconstrained subclass refers to methods that do not use any additional constraint conditions:

1. **LDA(L, H)**: This method is based on LDA, but uses labeled examples L as seeds for each group/topic. All labeled examples in L always stay in the same topic, which is referred to as hard initialization (H). The treatment of the labeled example set L is similar below.
2. **DF-LDA(L, H)**: DF-LDA is an LDA method [7] that incorporates must-link and cannot-link constraints. Our labeled example set L can be expressed as a combination of must-link and cannot-link constraints. Unfortunately, due to the huge number of cannot-link constraints, the system is prone to crashing, so only must-link constraints can be used. For example, in the car dataset, the number of

cannot-link constraints is 194,400 for 10% labeled data, and it goes up to 466,560,000 for 20% labeled data. DF-LDA also has a parameter η controlling the strength of the links, which we set to a very high value (=1000) to reflect hard initialization. We did not use DF-LDA in the above unsupervised subclass as without constraints, it simplifies to regular LDA.

3. **K-means(L, H)**: This method is based on K-means, but the clustering result of the labeled seeds is fixed and unchanged during the initialization stage.

4. **EM(L, H)**: This is the earliest proposed Expectation-Maximization (EM) algorithm in semisupervised learning. In this method, only labeled examples are used as initial seeds. The EM algorithm iteratively optimizes the model parameters to gradually improve the classification accuracy on unlabeled data. It utilizes the prior knowledge of labeled examples to guide the classification of unlabeled data and performs parameter estimation by maximizing the likelihood function.

For the hard-constraint (H) subclass (where we apply two constraints that cannot be violated), we have the following methods (LC is the combination of labeled (L) and semilabeled (SL) data generated by the constraints (C)):

1. **Rand(LC, H)**: This is an important baseline method. It demonstrates whether using constraints alone is sufficient to produce good results. In other words, the final result is the expanded seed SL combined with the remaining unlabeled data U, randomly assigned to different groups. This method can be used to evaluate the impact of constraints alone on the results, without considering the influence of other algorithms.

2. **LDA(LC, H)**: It is similar to LDA(L, H), but both the initial seed L and the expanded seed SL are considered as labeled examples. They also remain in the same topic/group throughout the process. It should be noted that although SL is referred to as a set of soft-labeled examples (seeds) in the proposed algorithm. By considering SL as hard-labeled examples, we can explore their influence on the classification results.

3. **DF-LDA(LC, H)**: This is DF-LDA with both L and SL represented as mandatory-link constraints. Similarly, a larger $\eta = 1000$ is used to ensure that the mandatory-link constraints of L and SL are not violated.

4. **K-means(LC, H)**: It is similar to K-means(L, H), but both L and SL are kept within their assigned clusters. In this method, LC is used as the seed to perform the K-means clustering algorithm, ensuring that the labeled examples L and expanded seed SL remain in their initially assigned clusters.

5. **EM(LC, H)**: It is similar to SC-EM, but SL is added to the labeled set L, and their categories are not allowed to change during the EM iterations. EM algorithm is a semisupervised learning method based on expectation-maximization, which improves the classification accuracy of unlabeled data through iterative optimization of model parameters. In this method, the expanded seed SL is added to the labeled set L to provide more prior information, and its labels are kept consistent by restricting their category from changing during EM iterations.

For the subclass of soft constraints, our two constraints can be violated. Initially, both the initial seed L and the extended seed SL are considered as labeled data, but later on, only L is considered as labeled data. The algorithm will reestimate the labels of each feature expression in SL to make them more consistent with the new constraint conditions. This subclass has the following methods:

1. **LDA(LC, S)**: Contrary to LDA (LC, H). It allows the SL set to change topics/ groups. In this approach, feature expressions in SL can switch to different topics or groups during the iteration process of the algorithm. By allowing the SL set to change topics/groups, this method can adapt more flexibly to changes in the data.
2. ***K*-means (LC, H)**: Contrary to Kmeans (LC, H). In this approach, we use LC as the seed to perform the *K*-means clustering algorithm, but unlike hard constraint subclasses, the extended seed SL is allowed to switch to different clusters during the clustering process. By allowing SL to switch between different clusters, this method can flexibly adapt to changes in the data while maintaining the initial assignment of clusters for L and SL.

Evaluation Results

We compared the results of SC-EM with 14 benchmark methods. To gain a deeper understanding of the impact of different numbers of labeled examples (seeds), we selected 10%, 20%, 30%, 40%, and 50% of the feature expressions from the gold standard data as the labeled set L, while the remaining portion was used as the unlabeled set U for experimentation. To ensure the reliability of the experimental results, we conducted 30 runs of the algorithm and reported their average results. However, due to space limitations, we can only present the detailed purity (Pur), entropy (Ent), and accuracy (Acc) results for using 30% labeled data (70% unlabeled data) in Table 3.4. For other proportions of labeled data, we summarized them in Table 3.5. Each result in Table 3.5 is calculated as the average value across five datasets. All the results are obtained from the unlabeled set U, which serves as our test set. Regarding the evaluation metrics, a lower value is better for entropy, while a higher value is better for purity and accuracy. In these experiments, we used a window size of $t = 5$. Through these experiments, our aim is to explore the influence of different numbers of labeled examples on algorithm performance in order to better understand and utilize our .

According to the clear results shown in Tables 3.4 and 3.5, the proposed algorithm (SC-EM) outperforms 14 benchmark methods on each dataset significantly. Specifically, the following observations can be made:

- LDA, mLSA, and Kmeans without seeds (labeled data) perform the worst. This is understandable as seeds are helpful for improving the results. Without seeds, DF-LDA performs the same as the traditional LDA method.
- LDA-based methods appear to be the weakest, while *K*-means based methods perform slightly better, but the EM-based method performs the best. This clearly

Table 3.4 Comparison results (L = 30% of the gold standard data)

Methods	Home theater			Insurance			Mattress			Car			Vacuum		
	Acc	Pur	Ent	Acc	Pur	Ent	Acc	Pur	Ent	Acc	Pur	Ent	Acc	Pur	Ent
LDA	0.06	0.31	2.54	0.11	0.36	2.24	0.05	0.32	2.57	0.06	0.37	2.39	0.03	0.36	2.09
mLSA	0.06	0.31	2.53	0.14	0.38	2.19	0.06	0.34	2.55	0.09	0.37	2.40	0.03	0.37	2.11
Kmeans	**0.21**	**0.42**	**2.14**	**0.25**	**0.45**	**1.90**	**0.15**	**0.39**	**2.32**	**0.25**	**0.44**	**2.16**	**0.24**	**0.47**	**1.78**
LDA(L, H)	0.10	0.32	2.50	0.16	0.37	2.22	0.10	0.34	2.57	0.19	0.39	2.36	0.10	0.39	2.09
DF-LDA(L, H)	0.27	0.37	2.32	0.25	0.41	2.00	0.19	0.39	2.35	0.28	0.45	2.15	0.31	0.40	1.98
Kmeans(L, H)	0.20	0.42	2.12	0.25	0.43	1.92	0.17	0.42	2.26	0.27	0.48	2.04	0.20	0.48	1.76
EM(L, H)	**0.48**	**0.50**	**1.93**	**0.50**	**0.53**	**1.69**	**0.52**	**0.56**	**1.87**	**0.56**	**0.58**	**1.80**	**0.49**	**0.52**	**1.79**
Rand(CL, H)	0.41	0.46	2.07	0.40	0.46	1.94	0.40	0.47	2.07	0.34	0.41	2.31	0.39	0.52	1.59
LDA(CL, H)	0.44	0.50	1.96	0.42	0.48	1.89	0.42	0.49	1.97	0.44	0.52	1.87	0.43	0.55	1.48
DF-LDA(CL, H)	0.35	0.49	1.86	0.33	0.49	1.71	0.23	0.39	2.26	0.34	0.51	1.88	0.37	0.52	1.58
Kmeans(CL, H)	0.49	0.55	1.70	0.48	0.55	1.62	0.44	0.51	1.91	0.47	0.54	1.80	0.44	0.58	1.42
EM(CL, H)	**0.59**	**0.60**	**1.62**	**0.58**	**0.60**	**1.46**	**0.56**	**0.59**	**1.74**	**0.62**	**0.64**	**1.54**	**0.55**	**0.60**	**1.44**
LDA(CL, S)	0.24	0.35	2.44	0.27	0.40	2.14	0.23	0.37	2.44	0.27	0.41	2.33	0.23	0.41	2.01
Kmeans(CL, S)	0.33	0.46	2.04	0.34	0.45	1.90	0.25	0.43	2.20	0.29	0.47	2.07	0.37	0.50	1.68
SC-EM	**0.67**	**0.68**	**1.30**	**0.66**	**0.68**	**1.18**	**0.68**	**0.70**	**1.27**	**0.70**	**0.71**	**1.24**	**0.67**	**0.68**	**1.18**

Table 3.5 Influence of the seeds' proportion (which reflects the size of the labeled set L)

Methods	Acc					Pur					Ent				
	10%	20%	30%	40%	50%	10%	20%	30%	40%	50%	10%	20%	30%	40%	50%
LDA	0.07	0.07	0.06	0.06	0.08	0.33	0.33	0.34	0.35	0.38	2.5	2.44	2.37	2.28	2.11
mLSA	0.07	0.07	0.08	0.07	0.07	0.34	0.35	0.35	0.37	0.38	2.48	2.42	2.36	2.26	2.12
Kmeans	**0.22**	**0.23**	**0.22**	**0.22**	**0.22**	**0.42**	**0.43**	**0.44**	**0.44**	**0.46**	**2.16**	**2.11**	**2.06**	**1.98**	**1.86**
LDA(L, H)	0.10	0.10	0.13	0.14	0.15	0.34	0.34	0.36	0.37	0.39	2.48	2.43	2.35	2.25	2.11
DF-LDA(L, H)	0.23	0.25	0.26	0.27	0.30	0.41	0.40	0.41	0.41	0.44	2.23	2.23	2.16	2.10	1.94
Kmeans(L, H)	0.13	0.16	0.22	0.24	0.28	0.42	0.43	0.45	0.45	0.48	2.15	2.11	2.01	1.95	1.79
EM(L, H)	**0.35**	**0.44**	**0.51**	**0.55**	**0.58**	**0.43**	**0.19**	**0.54**	**0.57**	**0.61**	**2.22**	**1.99**	**1.81**	**1.65**	**1.49**
Rand(CL, H)	0.28	0.35	0.39	0.42	0.45	0.39	0.43	0.47	0.50	0.54	2.33	2.15	2.00	1.82	1.63
LDA(CL, H)	0.31	0.38	0.43	0.46	0.49	0.43	0.47	0.51	0.54	0.58	2.16	1.99	1.83	1.69	1.49
DF-LDA(CL, H)	0.32	0.33	0.33	0.34	0.36	0.49	0.50	0.48	0.48	0.48	1.90	1.85	1.86	1.83	1.82
Kmeans(CL, H)	0.33	0.41	0.46	0.49	0.52	0.47	0.51	0.55	0.57	0.61	1.98	1.82	1.69	1.56	1.42
EM(CL, H)	**0.44**	**0.54**	**0.58**	**0.61**	**0.64**	**0.49**	**0.57**	**0.61**	**0.64**	**0.67**	**1.98**	**1.72**	**1.56**	**1.40**	**1.25**
LDA(CL, S)	0.17	0.21	0.25	0.30	0.34	0.34	0.36	0.39	0.42	0.46	2.47	2.37	2.27	2.09	1.87
Kmeans(CL, S)	0.23	0.28	0.32	0.36	0.42	0.43	0.44	0.46	0.48	0.51	2.15	2.08	1.98	1.86	1.70
SC-EM	**0.45**	**0.58**	**0.68**	**0.75**	**0.81**	**0.50**	**0.61**	**0.69**	**0.76**	**0.82**	**1.95**	**1.56**	**1.24**	**0.94**	**0.69**

indicates that the performance of classification (EM) is superior to clustering. Comparing the results of DF-LDA and *K*-means, they show similar performance.

- For LDA and Kmeans, the performance of hard-constraint methods (i.e., LDA(L, H) and Kmeans(L, H)) is better than that of soft-constraint methods (i.e., LDA (LC, S) and Kmeans(LC, S)). This suggests that the soft-constraint versions may incorrectly assign some correctly constrained expressions to other groups. However, for the EM-based method, the soft-constraint method (SC-EM) is significantly better than the hard-constraint version (EM (LC, H)). This indicates that the Bayesian classifier used in EM can leverage soft constraints and correct the erroneous assignments caused by constraints. In different settings, relying solely on constraints (such as synonyms and word sharing) is far from sufficient compared to SC-EM. EM can significantly improve the results.
- In the comparison of EM-based methods, we can see that SL (soft seeds) has a significant impact on all datasets. In particular, the SC-EM method is clearly the best.
- As the number of labeled examples increases from 10% to 50%, the results of each method improve (except for minor changes in DF-LDA). This means that increasing the number of labeled examples can enhance the algorithm's performance, making it better for classification or clustering.

Varying the Context Window Size

In order to investigate the impact of the text window size (t) on the performance of the SC-EM algorithm, we conducted a series of experiments by varying the value of t from 1 to 10. The results of these experiments are summarized in Fig. 3.5, which presents the average values across 5 datasets. In the evaluation results, we considered

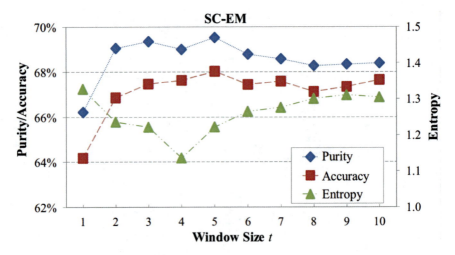

Fig. 3.5 Influence of context window size

three metrics: purity, accuracy, and entropy. For purity and accuracy, higher values are desirable, while for entropy, lower values are preferred.

From the graph, it can be clearly observed that the SC-EM algorithm achieves similar and good results when the window size is between 2 and 6. This indicates that within this range, the specific value of the window size has a minimal impact on the algorithm's performance. However, when the window size is less than 2 or greater than 6, there is a noticeable decrease or fluctuation in the results' performance.

It is important to emphasize that all the evaluation experiments were conducted with a text window size of $t = 5$. By observing the results obtained with different window sizes, we can gain a better understanding of the performance characteristics of the SC-EM algorithm under different parameter settings, providing valuable guidance for further optimization and adjustments.

3.2.6   Conclusion

This section introduces a feature grouping task in a semisupervised setting, aiming to address a problem that requires some form of supervision, because its solution depends on the user's application needs. In this task, we aim to effectively group relevant features to better understand the data and extract useful information for data analysis and machine learning tasks.

To address this problem, this section proposes a method using the EM algorithm and introduces two soft constraints to improve the algorithm's performance. The EM algorithm is an iterative optimization algorithm that gradually optimizes model parameters by alternating between an expectation step (E-step) and a maximization step (M-step) to obtain more accurate results. The introduction of soft constraints is to better constrain the correlations between features, improving the accuracy and stability of the grouping.

To evaluate the proposed method, empirical evaluations were conducted using five real datasets as benchmarks. The evaluation results demonstrate that the proposed method outperforms 14 baseline methods on various datasets. This indicates that the proposed method has significant advantages in the semisupervised feature grouping task and can better meet the user's application needs.

3.3   Constrained LDA for Grouping Product Features in Opinion Mining

3.3.1   Introduction

Topic modeling is an effective method for sentiment mining, which we employ in this section. Firstly, we extend a popular topic modeling method, namely the LDA method, to handle large-scale constraints. Next, we propose two new methods for automatically extracting two types of constraints. These constraints can further guide the topic modeling process and help accurately capture topics related to specific domains or sentiments. By introducing these constraints, we can have better control over the outcomes of topic modeling, making them more aligned with practical application needs. Finally, we apply the obtained constrained-LDA model and extracted constraints to group product features. By inputting product reviews or textual data into the constrained-LDA model, we can cluster and categorize similar topic words based on the relationship between constraints and topics, thereby achieving effective grouping of product features. Experimental results demonstrate that constrained-LDA outperforms the original LDA model and the latest mLSA method in terms of performance. This indicates that constrained-LDA has significant advantages and potential applications in sentiment mining and product feature grouping tasks.

3.3.2   The Proposed Algorithm

The original Latent Dirichlet Allocation (LDA) is a popular topic model widely used in unsupervised text analysis. It has been demonstrated in semisupervised clustering research [8, 9] that prior knowledge can guide the operation of clustering algorithms, resulting in more accurate and interpretable clustering results. Similarly, we believe that this knowledge can also help improve the performance of the LDA model, as LDA is essentially a clustering algorithm that assigns text data to different topics. In this section, we will first introduce the basic principles and operation of LDA to gain a deeper understanding of it. Next, we will provide a detailed explanation of the proposed constrained-LDA method, which utilizes prior knowledge to guide the training process of the topic model. By incorporating prior knowledge into the LDA model, we can infer topics more accurately and achieve effective grouping of product features. This constrained-LDA method, which combines prior knowledge, has significant application potential in tasks such as sentiment mining, topic analysis, and product feature grouping.

Introduction to LDA

Topic modeling is a method for processing textual documents. Unlike many other models, it treats each document as a "bag of topics" composed of different topics. The goal of topic modeling is to classify each word in a document into the appropriate topic. In this field, several probabilistic topic models have been proposed [7, 10–18], with LDA being one of the most popular methods. The input of an LDA model is a word × document matrix, and the output is the document-topic distribution θ and the topic-word distribution ϕ.

To obtain these distributions, researchers have proposed two main algorithms: the EM [11] and Gibbs sampling [19]. In our study, we chose to use the Gibbs sampling algorithm to infer the topic distribution of the documents. In the Gibbs sampling-based LDA model, the most crucial step is to update the topic for each word in each document based on the probabilities calculated using Eq. (3.19).

$$P(z_i = k \mid w_i = v, \mathbf{z}_{-i}, \mathbf{w}_{-i}) = \frac{C_{vk}^{WT} + \beta}{\sum_v' C_{vk}^{WT} + V\beta} \frac{C_{dk}^{DT} + \alpha}{\sum_{k'} C_{dk}^{DT} + K\alpha} \tag{3.19}$$

where $z_i = k$ represents the assignment of the ith word to topic k, indicating that the word is associated with topic k, $w_i = v$ signifies that the observed ith word corresponds to the vth word in the vocabulary of the text corpus, representing its position in the vocabulary, $\mathbf{z}_{-i}$ denotes the topic assignment for all words except the ith word, C_{vk}^{WT} indicates the number of times word v is assigned to topic k, that is, how many times word v is associated with topic k in the document, C_{dk}^{DT} represents the number of times topic k occurs in document d, indicating how many times topic k appears in document d. K is the number of topics, which is provided by the user, V is the size of the vocabulary, representing the total number of different words in the text corpus, α and β are hyperparameters that control the learning process of the topic model, α is the distribution parameter for document-topic distribution, and β is the distribution parameter for topic-word distribution. By default, they are set to $50/K$ and 0.01, respectively.

After N iterations of Gibbs sampling for all terms, we obtain two important distributions: the document-topic distribution θ and the topic-word distribution ϕ. Specifically, Eq. (3.20) is used to estimate the document-topic distribution θ, which informs us about the probability distribution of each topic within each document. On the other hand, Eq. (3.21) is used to estimate the topic-word distribution ϕ, which informs us about the probability distribution of different words within each topic.

$$\theta_{dk} = \frac{C_{dk}^{DT} + \alpha}{\sum_{k'} C_{dk}^{DT} + K\alpha} \tag{3.20}$$

$$\phi_{vk} = \frac{C_{vk}^{WT} + \beta}{\sum_{v'} C_{vk}^{WT} + V\beta} \qquad (3.21)$$

Constrained-LDA

Constrained-LDA is an improved approach that introduces constraints to traditional LDA (Latent Dirichlet Allocation). These constraints are derived from existing knowledge, assuming that each word in the constraint can only belong to one topic. In contrast to LDA, constrained-LDA requires two additional inputs: a set of must-link constraints and a set of cannot-link constraints. Must-link constraints specify that two words should belong to the same topic, while cannot-link constraints specify that two words should not belong to the same topic.

The main idea of the proposed method is to use the probabilistic information from the constraints to adjust the topic update probabilities obtained from LDA calculations. During the topic update process (as shown in Eq. (3.19)), for each candidate topic k ($k \in \{1, 2, \ldots, K\}$), we calculate additional probabilities, $q(z_i = k)$, based on the must-link and cannot-link constraints. These additional probabilities are then multiplied with the probabilities obtained from the original LDA model to obtain the final probabilities used for topic updates. Please refer to Eq.(3.22) for more detailed information.

$$P(z_i = k \mid w_i = v, \mathbf{z}_{-i}, \mathbf{w}_{-i}) = q(z_i = k) \frac{C_{vk}^{WT} + \beta}{\sum_{v'} C_{vk}^{WT} + V\beta} \frac{C_{dk}^{DT} + \alpha}{\sum_{k'} C_{dk}^{DT} + K\alpha} \qquad (3.22)$$

As illustrated by Eqs. (3.19) and (3.22),$q(z_i = k)$ plays a key role in constrained-LDA, because $q(z_i = k)$ represents intervention or help from pre-existing knowledge of must-links and cannot-links. In this study, $q(z_i = k)$ is computed as follows: For the given term w_i, if w_i is not constrained by any must-links or cannot-links, $\{q(z_i = k) \mid k = 1, \ldots, K\} = 1$; otherwise, $\{q(z_i = k) \mid k = 1, \ldots, K\}$ is calculated using the following 4 steps in Fig. 3.6 and Algorithm 3.6.

1. **Step 1**: We want to obtain the weights of the mandatory and forbidden topics for the vocabulary term w_i. Mandatory topics refer to the topics to which the vocabulary term w_i should be assigned, providing topic information related to w_i. Forbidden topics, on the other hand, are the topics to which the vocabulary term w_i should not be assigned, helping to exclude topics unrelated to w_i.

   Specifically, we first query the mandatory and forbidden repositories to find the mandatory and forbidden terms associated with the vocabulary term w_i. Through the query, we can obtain lists of terms that are mandatory and forbidden for w_i. Next, we further obtain the topic information for these relevant terms from the topic modeling. This can be achieved by inputting the relevant terms into a topic model such as LDA (Latent Dirichlet Allocation). The topic model will

Fig. 3.6 Computing the
weights for must-topics and
cannot-topics

	Step 1		Step 2
	$M_1[1]$	$T_1(2)$	$T_1(2 \times \lambda)$
must	...	...	...
	$M_i[K]$	$T_K(9)$	$T_K(9 \times \lambda)$
	$C_1[1]$	$t_1(3)$	$t_1(3 \times (1-\lambda))$
cannot	...	...	...
	$C_j[K]$	$t_K(6)$	$t_K(6 \times (1-\lambda))$

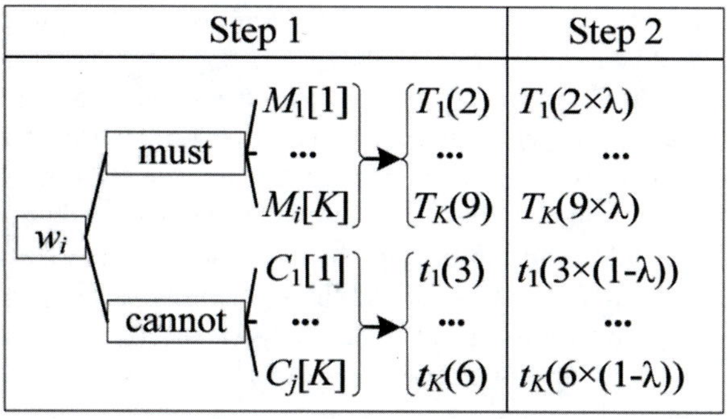

assign one or more topics to each term. Therefore, we can obtain the weights of the mandatory and forbidden topics for the vocabulary term w_i.

To illustrate with an example, let's assume that the mandatory terms for the vocabulary term w_i are M_1 and M_2, and the forbidden terms are C_1, C_2, C_3. Then, we use a topic model (e.g., LDA) to assign M_1, M_2 and C_1, C_2, C_3 to a topic k. This means that M_1 and M_2 are assigned to the mandatory category of topic k, while C_1, C_2, and C_3 are excluded from topic k. Therefore, for topic k, the weight of the mandatory topic for the vocabulary term w_i is weight($w_i, T_k(|\{M_1, M_2\}|)$) =weight($w_i, T_k(2)$) = 2, indicating that 2 relevant terms are assigned to the mandatory category of topic k. At the same time, the weight of the forbidden topic for the vocabulary term w_i is weight($w_i, t_k(|\{C_1, C_2, C_3\}|)$)=weight($w_i, t_k(3)$) = 3, indicating that 3 relevant terms are assigned to the forbidden category of topic k. Here, weight(w_i, T_k) or weight(w_i, t_k) represents whether the vocabulary term w_i should or should not be assigned to topic k.

Using the weights of the mandatory and forbidden topics obtained in step 1, we can utilize this information in subsequent steps to update and adjust the topics, thereby improving the performance and accuracy of the model.

2. **Step 2**: Step 2 involves adjusting the relative influences between the must-link and cannot-link categories. When extracting these two types of constraints, the qualities of must-links and cannot-links may differ. To account for this, we introduce a damping factor λ to adjust their relative influences. Specifically, all the weights of the must-link topics are multiplied by λ, while the weights of the cannot-link topics are multiplied by $(1 - \lambda)$.

For instance, if $T_k(2)$ is a must-link topic, its weight will be adjusted to $T_k(2 \times \lambda)$, while if $t_k(3)$ is a cannot-link topic, its weight will be adjusted to $t_k(3 \times (1 - \lambda))$. In this study, we empirically set the default value of λ to 0.3.

Based on the results of the previous two steps, we further propose Steps 3 and 4, which aim to convert the weights of the must-link and cannot-link topics to $\{q(z_i = k) \mid k = 1, \ldots, K\}$, as outlined in Algorithm 3.6.

3. **Step 3**: Step 3 aims to aggregate the weights for each candidate topic. For a given term w_i, it can be categorized into three types of candidate topics: must-topics, unconstrained topics, and cannot-topics. Must-topics represent the topics to which w_i should be assigned, while cannot-topics represent the topics to which

w_i should not be assigned. Therefore, we need to calculate the probability that w_i will be assigned to a candidate topic, k.

During the calculation, if candidate topic k belongs to the must-topics, we add the weight (w_i, T_k) to $q(z_i = k)$ to increase the probability of assigning w_i to topic k. This is done to emphasize the relevance between w_i and topic k, thereby increasing the likelihood of w_i being assigned to that topic. Conversely, if candidate topic k belongs to the cannot-topics, we subtract the weight (w_i, t_k) from $q(z_i = k)$ to decrease the probability of assigning w_i to that topic. This is done to weaken the relevance between w_i and topic k, thereby reducing the chance of w_i being mistakenly assigned to that topic. The specific operations can be referred to lines 2–6 in Algorithm 3.6.

By performing Step 3, we are able to aggregate the weights of term w_i across different candidate topics, obtaining the distribution probability of w_i over the topics. Such results help us understand the association between terms and topics, further enhancing the accuracy of topic assignment.

Algorithm 3.6: Probability Aggregation and Relaxation

Input:	w_i;
	w_i 's must-topics' weights: weight (w_i, T_k), $k = 1, 2, \ldots, K$;
	w_i 's cannot-topics' weights: weight (w_i, t_k), $k = 1, 2, \ldots, K$;

Output: $\{q(z_i = k) \mid k = 1, 2, \ldots, K\}$
1:   Initial all $\{q(z_i = k) \mid k = 1, 2, \ldots, K\}$ to zero
2:   //**Step 3** - Aggregate
3:   **for** $(k$ **in** $\{1, 2, \ldots, K\})$
4:      **if** $(k$ **in** $\{w_i$'s must-topics $\}) q(z_i = k) + =$ weight (w_i, T_k)
5:      **if** $(k$ **in** $\{w_i$'s cannot-topics $\}) q(z_i = k) - =$ weight(w_i, t_k)
6:
7:   //**Step 4** - Normalize and relax
8:   max $= \{q(z_i = k) \mid k = 1, 2, \ldots, K\}_{\max}$
9:   min $= \{q(z_i = k) \mid k = 1, 2, \ldots, K\}_{\min}$
10:   **for** $(k$ **in** $\{1, 2, \ldots, K\})$
11:      $q(z_i = k) = \frac{q(z_i = k) - \min}{\max - \min}$
12:      $q(z_i = k) = q(z_i = k) \times \eta + (1 - \eta)$

In the above example, for the candidate topic k, the weight $q(z_i = k)$ is: $0+$ weight $(w_i, T_k(2 \times \lambda))$ - weight $(w_i, t_k(3 \times (1 - \lambda))) = 2 \times \lambda - 3 \times (1 - \lambda) = 5\lambda - 3$.

4. **Step 4**: Normalize and relax the weights of each candidate topic. In practical applications, constraints, especially automatically extracted constraints, cannot guarantee complete accuracy. Therefore, we need a parameter to adjust their influence on the model based on the quality of the constraints. When the constraints are highly reliable and accurate, we want the model to treat them as hard constraints, meaning they must be strictly followed. However, when there are errors or uncertainties in the constraints, we need to relax them to allow the model to partially disregard these constraints.

To achieve this goal, we introduce a relaxation factor η to adjust the weights of $\{q(z_i = k) \mid k = 1, \ldots, K\}$. Specifically, when the constraints are highly reliable, the value of η should be close to 1 to maintain the strength of the constraints. When there are uncertainties or errors in the constraints, the value of η should be close to 0 to weaken the impact of the constraints.

By normalizing and relaxing $\{q(z_i = k) \mid k = 1, \ldots, K\}$, we are able to maintain the flexibility of the model while considering the influence of the constraints to a reasonable extent. This approach helps improve the robustness of the model, enabling it to generate reasonable results even in the presence of incomplete and inaccurate constraints. For specific adjustment methods, refer to Algorithm 3.6.

Before being relaxed, $\{q(z_i = k) \mid k = 1, \ldots, K\}$ are normalized to $[0, 1]$ using Eq. (3.23) (lines 8–11 in Algorithm 3.6). In Eq. (3.23), max and min represent the maximum and minimum values of $\{q(z_i = k) \mid k = 1, \ldots, K\}$, respectively.

$$q(z_i = k) = \frac{q(z_i = k) - m}{m - m} \tag{3.23}$$

Then, $\{q(z_i = k) \mid k = 1, \ldots, K\}$ are relaxed by the relaxation factor η based on Eq. (3.24) (line 12 in Algorithm 3.6). The default value of η is set to 0.9 in our study.

$$q(z_i = k) = q(z_i = k) \times \eta + (1 - \eta) \tag{3.24}$$

In our application of grouping product features, please take note of the following points. Firstly, we consider each product feature as a term. We can estimate the parameter ϕ by using Eq. (3.21), which is used to describe the relationship between topics and terms. By applying this equation, we can identify a set of topics, where each topic consists of a set of related terms. Such output results can help reveal the intrinsic connections and commonalities among product features.

3.3.3   Constraint Extraction

We now come back to our application and discuss how to extract constraints automatically. The general idea has been discussed earlier. For completeness, we briefly discuss them here again When we extract constraint conditions automatically in our application, there are two types of constraints: Must-link and Cannot-link.

Must-link

Must-link refers to the situation where two product features share one or more words, indicating that they should belong to the same theme. For example, if we have two product features, "battery power" and "battery life," and they both contain the word "battery," they form a mandatory constraint, indicating that they belong to the same theme. If two product features f_i and f_j share one or more words, we assume them to form a must-link, that is, they should be in the same topic, for example, "battery power" and "battery life." Clearly, this method is not perfect. Then, the constraint relaxation mechanism comes to help.

Cannot-link

Cannot-link refers to the situation where two product features appear in the same sentence without being connected by "and." In such cases, they should be considered as not belonging to the same theme. This is because people typically do not repeat the same feature within the same sentence. However, if the features are connected by "and," we cannot be certain whether they belong to the same theme, because using "and" sometimes introduces uncertainty. Generally, features based on product names are more likely to be connected together.

These constraint conditions help us better understand and organize product features in our application, thereby improving the accuracy and efficiency of the automation process.

3.3.4 Experiments

In this section, we will test and evaluate a newly proposed constrained-LDA model to address a specific problem. We will conduct tests on this model in different scenarios and compare it with the original LDA algorithm and the recently popular multilevel mLSA method. We have chosen not to compare it with the similarity-based method in [14] because their technique requires a predetermined feature taxonomy, which we did not use in this study.

Data Sets

To demonstrate the applicability of our proposed algorithm in different domains, we conducted experiments in two fields: digital cameras and smartphones. We used two datasets that contained customer reviews and have been widely used in opinion mining research. To increase the size of the datasets, we obtained numerous additional reviews for other cameras and smartphones from Amazon. We performed detailed feature annotation for each dataset to be used in our system. For specific information about the datasets, please refer to Table 3.6.

Table 3.6 Summary of the data sets

Camera	Number of reviews	2400	Phone	Number of reviews	1315
	Number of sentences	20,628		Number of sentences	18,393
	Number of vocabulary	7620		Number of vocabulary	7376

Gold Standard

In this study, we utilized datasets of customer reviews that included comments on digital cameras and smartphones. To gain a better understanding of the product features mentioned in these reviews, we enlisted the help of experts to annotate these features. For the digital camera dataset, we categorized the features into 14 themes based on a camera classification system introduced by a product called Active Sales AssistantTM. This product is a renowned guided selling solution offered by Active Decisions company. If you would like to obtain further information, you can visit their website at www.activebuyersguide.com [15]. As for the smartphone dataset, we employed several themes introduced by Google product releases to classify the features of the smartphones, ultimately dividing them into 9 themes.

Evaluation Measure

The Rand Index [16] is a measure used to evaluate the performance of our product feature grouping algorithm and has been utilized by many researchers [8, 17, 18]. It is also an evaluation metric mentioned in another article. we will compare it with another method called mLSA. The Rand Index is employed to assess the consistency between two partitions, "P_{answer} and $P_{machine}$," of the same dataset D [20]. Each partition can be viewed as a collection of pairwise decisions, where n represents the size of the dataset D. For any two instances, I_j and I_k, in the partition, P_{answer} and $P_{machine}$ either assign them to the same cluster or to different clusters. We denote a as the number of correct decisions where I_j and I_k are assigned to the same cluster in both P_{answer} and $P_{machine}$ and b as the number of correct decisions where these two instances are assigned to different clusters in the two partitions. The overall consistency can be calculated using the Eq. (3.25). In our study, all product features constitute the dataset D; P_{answer} represents our gold standard, and $P_{machine}$ represents our experimental results.

$$\mathbf{RI}(P_{answer}, P_{machine}) = \frac{a + b}{C_n^2} = \frac{a + b}{n \times (n - 1)/2} \tag{3.25}$$

Compared with LDA

References [7] proposed the latest LDA model called DF-LDA, which can consider certain special association and exclusion relationships. However, DF-LDA is unable to handle many such relationships. When we only use one-fifth of the relationships in Table 3.7, DF-LDA causes system crashes. Therefore, we cannot use the DF-LDA model for tasks that require grouping a large number of product features. Due to the limitations of DF-LDA, we can only report comparative results with the original LDA model.

The original LDA and constrained LDA are two methods used in different domains with varying numbers of topics. The numbers of topics used are 20, 40, 60, 80, 100, and 120, respectively. It is important to note that LDA requires the user to determine the number of topics beforehand. Additionally, we did not report the results using the original topic numbers (14 and 9) because they performed poorly (as seen in Fig. 3.7). We conducted experiments using only mandatory constraints, only prohibited constraints, and their combinations. For each dataset, we extracted a certain number of unique constraint pairs (in our case, pairs (a, b) and (b, a) were considered the same). All experimental results are shown in Fig. 3.7. From Fig. 3.7, it can be observed that the patterns of different methods are generally consistent across different datasets, indicating consistent results. Next, we present some additional observations:

- The constrained methods (LDA + cannot, LDA + must, and LDA + must + cannot) outperform the original LDA model in terms of performance. When the number of topics is small, there is an improvement of over 10% in the digital camera corpus and approximately 7% in the cell phone corpus. As the number of topics increases, the improvement slightly decreases but still remains at 7% for the digital camera corpus and 4% for the cell phone corpus. In other words, the constrained methods demonstrate superior performance in topic modeling tasks, particularly when dealing with a limited number of topics.
- Both the approach of not linking (LDA + cannot) and the approach of must-linking (LDA + must) perform well, although overall, not linking is slightly more effective. This phenomenon suggests that our assumption about not linking is reasonable, and the extracted not-linked parts are of higher quality. When the number of topics is small or large, must-linking performs slightly better. We believe the reason is that in these two extreme cases, either the not-linked words are forced to be assigned to the same topic (for a smaller number of topics) or they tend to be dispersed into too many topics. The original LDA model also exhibits this behavior, which is quite understandable.

Table 3.7 Number of the extracted constraints

Camera	Number of must-links	300	Phone	Number of must-links	184
	Number of cannot-links	5172		Number of cannot-links	5009

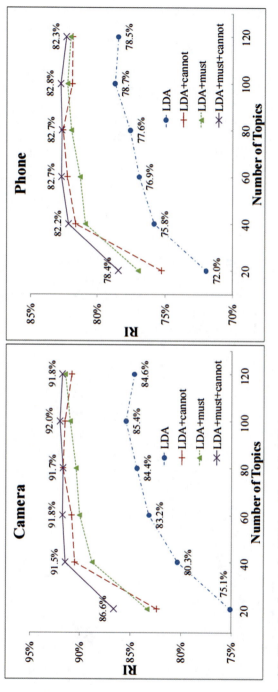

Fig. 3.7 RI results of constrained-LDA and the original LDA

- Combining the must-link constraint and the cannot-link constraint together (LDA + must + cannot) always yields better results than using either constraint alone (LDA + cannot or LDA + must). Although the degree of improvement is not significant, it is consistent. This also indicates that using the must-link and cannot-link constraints separately is already quite effective. In simple terms, applying both constraints simultaneously leads to better outcomes, while using either constraint alone also brings some benefits.
- In practical applications, it is generally more effective to use a smaller number of topics, as it allows users to understand and handle them more easily. In both cases, it seems that choosing 40 topics is the most suitable option. In other words, representing information with fewer topics can better assist people in understanding and dealing with it, and in these scenarios, 40 topics appear to be the optimal choice.

Comparing with mLSA

A recently proposed multilevel latent semantic association method, called mLSA [20], also solves the same problem as we do. We chose to compare our proposed constrained-LDA method with mLSA, rather than other existing methods. The comparison was based on both the digital camera corpus and the cell phone corpus, and the results were shown in Fig. 3.8. In the comparison, we only used 40 topics, which appeared to be the optimal number among our tested topic numbers in

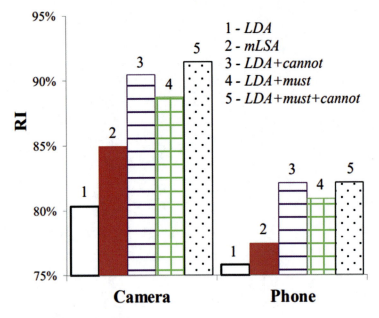

Fig. 3.8 Comparisons with mLSA

Fig. 3.7. The results showed that for the problem we studied, the performance of the constrained-LDA method is comparable to that of the mLSA method, and it is even better than other existing methods.

In Fig. 3.8, the results show the positive impact of constraints on the results. We can see that the mLSA method (red bar) achieved encouraging results by transforming the input document content before applying LDA. In contrast, our constrained-LDA model did not make any efforts to reorganize or transform the input document content, and our input was the set of original reviews. However, we found that the results produced by the constrained-LDA method (LDA + cannot, LDA + must, and LDA + must + cannot) are all substantially better than those of the mLSA method. The success of this method demonstrates that constraints can improve the efficiency and accuracy of topic models.

Influence of Parameters

Compared to the original LDA model, the proposed constrained-LDA model introduces two additional parameters, namely, damping factor λ and relaxation factor η. In this section, we will investigate the impacts of these two parameters on the overall performance of the model. The introduction of these parameters can make the model more flexible and better suited for different datasets and application scenarios. We will conduct experiments to study how these parameters affect the performance of the model, and provide more guidance and recommendations for using the constrained-LDA model in research and practice.

1. **Influence of the damping factor** $-\lambda$: The damping factor λ is used to control the relative influence of must-links and cannot-links in the constrained model. In Fig. 3.9, λ values from 0 to 1 indicate that the influence of cannot-links decreases while the influence of must-links increases. The performance of constrained-LDA slightly improves with increased influence of must-links over cannot-links, but drops sharply to the lowest point when only must-links are used ($\lambda = 1$). This illustrates the synergistic effect of must-links and cannot-links, as they help each other. Since results generated with λ values above 0.3 are very similar, we use $\lambda = 0.3$ as the default value. The default damping factor of $\lambda = 0.3$ was used in the experiments shown in Figs. 3.7 and 3.8.

2. **Influence of the relaxation factor** $-\eta$: In this study, the relaxation factor η represents the strength of constraints on the LDA model. When $\eta = 0$, the constrained-LDA model reduces to the original LDA model without any constraints. When $\eta = 1$, both must-link and cannot-link constraints become hard constraints that cannot be violated. Figure 3.10 shows the influence of η on the overall performance. The performance of LDA + cannot, LDA + must, and LDA + must + cannot significantly improves with the increase of constraint strength. This observation not only demonstrates the clear benefit of constraints for topic modeling (or LDA), but also indicates that the extracted must-links and cannot-links are of high quality, particularly for the extracted cannot-links. In

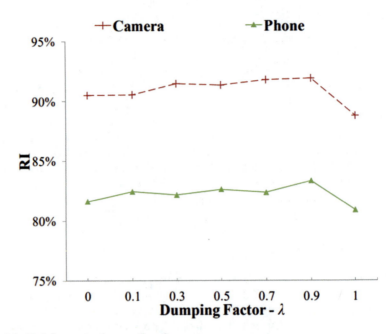

Fig. 3.9 λ's influence on the overall performance

fact, the best results on the digital camera and mobile phone data are obtained when using both must-links and cannot-links constraints simultaneously ($\eta = 1$). Since high-quality constraints are usually difficult to extract, we set the default value of η to 0.9 in our system. The default value of $\eta = 0.9$ was also used in the experiments reported in Figs. 3.7, 3.8, and 3.9.

3.3.5 Conclusion

This section proposes a method called Constrained-LDA to enhance the widely used topic modeling method LDA, allowing it to consider prior knowledge in the form of must-link and cannot-link constraints. The resulting model is highly flexible, with the ability to adjust the strength of constraints and the relative influence of must-links and cannot-links, making it suitable for various applications. Experimental results demonstrate that the chosen default values perform well. We use Constrained-LDA to group product feature synonyms in two opinion mining datasets, and the results significantly outperform existing methods, indicating that constraints as prior knowledge can help with unsupervised topic modeling. Additionally, this section proposes two automatic methods for extracting the two types of constraints, and their quality is verified.

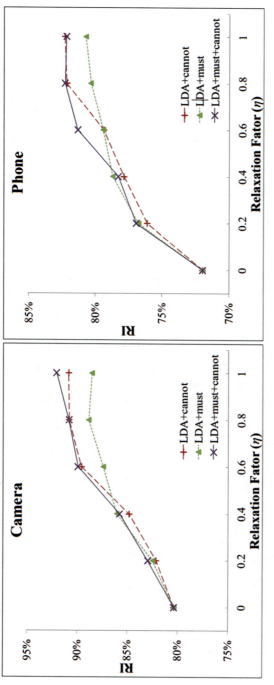

Fig. 3.10 η's influence on the overall performance (#topics = 40)

3.4   Product Feature Grouping for Opinion Mining

3.4.1   Introduction

In opinion mining tasks, people often use different words or phrases to describe the same product feature, such as "picture" and "photo" in the context of cameras, which actually refer to the same thing. To better summarize various opinions, it is necessary to group these synonyms. However, manual grouping is time-consuming and often involves many words. Therefore, we propose a more efficient method that can automatically group synonyms together. Our method mainly clusters words and phrases based on their similarities. Similarity measurement can be based on pre-existing knowledge resources, such as dictionaries or WordNet, or on the distributional properties of words in the corpus. We evaluated our method on reviews from five different domains and the results showed that our method is more effective than existing methods.

3.4.2   The Proposed Soft-Constrained Algorithm

The input to the proposed algorithm includes a set of reviews R and a set of feature expressions F extracted from R. Users can assign some of the feature expressions to predefined feature groups C as needed. Then, the algorithm uses some equations L to determine which feature expressions U should be assigned to which feature groups C. This algorithm is called the soft-constrained EM algorithm SC-EM, as shown in Algorithm 3.7. The algorithm relies on the following equations:

$$P(w_t|c_j) = \frac{1 + \sum_{i=1}^{|D|} N_{ti} P(c_j|d_i)}{|V| + \sum_{m=1}^{|V|} \sum_{i=1}^{|D|} N_{mi} P(c_j|d_i)} \tag{3.26}$$

$$P(c_j) = \frac{1 + \sum_{i=1}^{|D|} P(c_j|d_i)}{|C| + |D|} \tag{3.27}$$

$$P(c_j|d_i) = \frac{P(c_j) \prod_{k=1}^{|d_i|} P(w_{d_i,k}|c_j)}{\sum_{r=1}^{|C|} P(c_r) \prod_{k=1}^{|d_i|} P(w_{d_i,k}|c_r)} \tag{3.28}$$

In a training sample set D, each sample d_i is considered as a list of words. In Eq. (3.28), $w_{d_i,k}$ represents the ith word in d_i, and each word comes from the vocabulary $V = \{w_1, w_2, \cdots, w_{|V|}\}$. $C = \{C_1, C_2, \cdots, C_{|C|}\}$ is a set of predefined categories or groups, and N_{ti} represents the number of times the word w_t appears in d_i. To address the domain dependence issue mentioned earlier, we extract the distributional contexts of each feature expression for SC-EM. In simple terms, each example in L (or U) is actually the distributional context of the corresponding feature expression.

Algorithm 3.7: The Proposed Soft-Constrained Expectation Maximization Algorithm. The Algorithm Assigns Labels to a Set of Unlabeled Feature Expressions

---

Input: Labeled examples L;Unlabeled examples U
Output: Class labels of U
1:   Extract SL from U using constraints
2:   Learn an initial naive Bayesian classifier f
3:   using $L \bigcup SL$ and Eqs. (3.26) and (3.27)
4:   **repeat**
5:       //E-Step
6:       for each example d_i in U (including SL):
7:           Compute $P(c_j| d_i)$ using f and Eq. (3.28)
8:       //M-Step
9:       Learn a new naive Bayesian classifier f from $L \bigcup U$
10:      by computing $P(w_t| c_j)$ and $P(c_j)$ using Eqs. (3.26) and (3.27)
11: **until** the classifier parameters stabilize
12: Classify examples in U to C using the final classifier f

The EM algorithm suffers from the problem of local optima, so we use the constraints of natural language to provide better initialization by adding more possibly correct seeds, called the soft-labeled example set SL, which is mentioned in line 1 of Algorithm 3.7. For example, if the user-defined labeled example set L_i includes "appearance, appearance design," then the soft-labeled example set SL_i can include "appearance, design, exterior accessories." Since two soft constraints and lexical similarity are used to extract the soft-labeled example set SL, its correctness cannot be guaranteed completely. Therefore, during the learning process, we allow the soft-labeled example set SL to be violated, as shown in lines 2–11. Compared to the original EM algorithm, SC-EM has two main differences:

- In the first iteration (line 2), a set of soft-labeled example sets SL is applied in addition to the labeled example set L to initialize EM. Increasing the size of the training set can help produce better results.
- In the first iteration (line 2), the algorithm treats the examples in the soft-labeled example set SL equally as those in the labeled example set L, as training data for the initial classifier f_0. However, in the subsequent iterations (lines 4–11), the soft-labeled example set SL is treated as any example in the unlabeled example set U, and each iteration's classifier f_x predicts the unlabeled example set U. Then, a new classifier is constructed using the labeled example set L and the unlabeled example set U_{PL} with probability labels. This means that the class labels of the examples in the soft-labeled example set SL can change.

Finally, in line 12, the algorithm classifies the unlabeled example set U. Continuing with the example above, "exterior accessories" will be revised and removed from C_i by $SC - EM$ because its distributional context is different from other feature expressions in the labeled example set L_i and soft-labeled example set SL_i. On the other hand, some unlabeled feature expressions will be classified into C_i, such as "style, fashion," determined by the trained classifier.

3.4.3 Extracting the Example Set Using Constraints

Algorithm 3.8: The Algorithm for Generating the Soft-Labeled Set SL. The Resulting "Soft Seeds" Will Provide a Better Initialization for $SC - EM$

1: Construct a graph G with all feature expressions in U as nodes
2: **for** each feature expressions $u_m \in U$ **do**
3:     **for** each feature expressions $u_n \in U \neq$ **do**
4:         **if** $m \neq n$ & **isConstrained**(u_m, u_n) **then**
5:         add $edge(u_m, u_n)$ to G
6: Gather all connected components of G as CC
7: **for** each connected component $cc \in CC$ **do**
8:     **for** each feature group $L_i \in L$ **do**
9:         $score(L_i) \leftarrow 0$
10:     **for** each feature expression $e_1 \in cc$ **do**
11:        **for** each feature expression $e_2 \in L_i$ **do**
12:            **if** **isConstrained**(e_1, e_2) **then**
13:                $score(L_i) \leftarrow score(L_i) + Sim(cc, L_i)$
14:                **goto** the loop in line 8
15:        cc is added to SL_i such that $argmax_{L_j} score(L_i)$
16:     // Function for checking constraints
17:     **isConstrained**(u_m, u_n)
18:         **if** e_1 is a single word expression **then**
19:             **if** e_2 is a single word expression **then**
20:                 **if** e_1 and e_2 are synonyms **then**
21:                     **return** true
22:             **else** //e_2 is a phrase
23:                 **if** $e_1 \in e_2$ **then**
24:                     **return** true
25:         **else** // e_1 is a phrase
26:             **if** e_2 is a single-word expression **then**
27:                 **if** $e_2 \in e_1$ **then**
28:                     **return** true
29:             **else** // e_2 is a phrase
30:                 $s \leftarrow e_1 \bigcap e_2$;
31:                 **if** $|s| > 0$ **then**
32:                     **return** true
33:         **return** false
34:     // Function for calculating similarity between sets
35:     $Sim(cc, L_i)$:
36:         **return** $Avg_{e_1 \in cc, e_2 \in L_i}(\textbf{PrsSim}(p_1, p_2))$
37:     // Function for calculating similarity between phrases
38:     **PrsSim**(p_1, p_2):
39:         **return** $Max_{w_k \in p_1, w_q \in p_2} Jcn(w_k, w_q)$

Algorithm 3.8 is a detailed algorithm for extracting the soft seed set *SL*, which aims to provide better initialization for SC-EM. An important component of this algorithm is the subfunction is constrained(e_1, e_2), which is designed to check whether two feature expressions e_1 and e_2 are affected by the proposed constraints. To match e_1 and e_2, this function uses multiple matching methods. If both e_1 and e_2 are single words, the function checks whether they are synonyms. If either e_1 or e_2 is a phrase, or if both of them are feature expression phrases, the function checks whether they have shared words.

In the main part of the algorithm (lines 1–15), *L* is a collection of multiple subsets, that is, $L = \{L_1, L_2, \cdots, L_{|C|}\}$, where each subset L_i contains labeled examples (feature expressions) of the same category (same feature group). Similarly, SL is also a collection of multiple subsets, $SL = \{SL_1, SL_2, \cdots, SL_{|C|}\}$, where each subset SL_i contains soft-labeled examples (feature expressions) of the same category i (same feature group). Each subset in *L* and SL corresponds to the original labeled examples and the newly soft-labeled examples, respectively. These collections play an important role in the algorithm, used to extract and generate soft seed sets, to improve the initialization effect of SC-EM.

To extract SL_i from U using L_i, it is necessary to construct a graph (line 1) based on the feature expressions, where the feature expressions in U serve as the nodes. If the isConstrained function (lines 2–5) binds two feature expressions u_m and u_n, the algorithm connects them with an edge. According to experimental results, this graph *G* is usually very sparse. Then, a set of connected components CC can be collected from this graph without cutting any edges. Each connected component contains two or more members that may belong to the same feature group.

Next, we compare each connected component cc in CC with the labeled subsets $\{L_1, L_2, \cdots, L_{|C|}\}$ using WordNet-based lexical similarity (lines 7–15). We use connected components as the basic unit of comparison instead of individual feature expressions because some words may not be present in WordNet. For instance, when all the members of L_i and u_m are present in WordNet but u_n is not, we cannot use WordNet to compute the similarity between L_i and u_n. However, if u_m and u_n are connected through constraints, the similarity between um and L_i can be approximated as the similarity between un and L_i. As a result, we can consider un even if it is not present in WordNet.

Intuitively, if any members of *cc* have constraints with any members of L_i, it means that cc is likely to belong to L_i and can potentially be added to SL_i. However, conflicts may arise when cc is constrained with multiple labeled subsets L_i, and the question becomes which group cc should be assigned to. To resolve conflicts, we use a score to record the accuracy of grouping. After comparing cc with each labeled group L_i, the accumulated score is used to determine which class L_i has the strongest association with cc. We assign the class j with the highest score to *cc* and add all members of cc to SL_j. We use WordNet-based lexical similarity to calculate the score, with the Jcn algorithm performing the best among several options. These similarity measures rely on least common subsumer (LCS), the shared ancestor of two concept words. For example, the LCS of automobile and scooter is vehicle.

This study utilizes a WordNet-based lexical similarity calculation method to compute score values. The pairwise score, $\text{Sim}(cc, L_i)$, represents the average value of all cross-similarities among the members of cc and L_i, and it is computed by $\text{PrsSim}(p_1, p_2)$, where $p_1 \in cc$ and $p_2 \in L_i$. $\text{PrsSim}(p_1, p_2)$ calculates the WordNet similarity between two phrases, and the WordNet-based algorithm $\text{Jcn}(w_k, w_q)$ mentioned in line 39 is proposed by Jay Jiang and David Conrath [21]. Other similarity measures such as Res and Lin [22, 23] were also attempted, but Jcn performed the best for the given task. All of these similarity measures rely on varying degrees of least common subsumer (LCS), which is the shared ancestor of two concept words. For instance, "vehicle" is the LCS of "automobile" and "scooter." The similarity measure Res uses the information content of $\text{LCS}(w_1, w_2)$ as the similarity value, which is shown by two equations:

$$\text{Res}(w_1, w_2) = \text{IC}(\text{LCS}(w_1, w_2)) \tag{3.29}$$

$$\text{IC}(w) = -\log\text{Pr}(w) \tag{3.30}$$

where $\text{Pr}(w)$ is the probability of the word w based on the observed frequency counts in the WordNet corpus. Both Lin and Jcn try to refine Res by augmenting it with the information content of w_1 and w_2 using the following two equations:

$$\text{LCS}(w_1, w_2) = \frac{2 \times \text{Res}(w_1, w_2)}{\text{IC}(w_1) + \text{IC}(w_2)} \tag{3.31}$$

$$\text{Jcn}(w_1, w_2) = \frac{1}{\text{IC}(w_1) + \text{IC}(w_2) - 2 \times \text{Res}(w_1, w_2)} \tag{3.32}$$

3.4.4 Distributional Context Extraction

To apply the SC-EM algorithm, an example (bag of words) d_i needs to be provided for each feature expression e_i in naïve Bayesian learning. The example d_i is the aggregation of the distributional contexts of all sentences in L (or U) that contain the expression e_i. In a sentence s_{ij}, the context for e_i is the surrounding words of e_i within a text window of $[-t, t]$, including the words in e_i. Stop words are removed.

For example, e_i might be "screen," with two related sentences:

- $s_{i1} = $ "The LCD screen gives clear picture."
- $s_{i2} = $ "The picture on the screen is blurry."

We use the window size of $[-3\text{--}3]$. In that case, s_{i1} gives us $d_{i1} = \langle \text{the, LCD, screen, gives, clear, picture} \rangle$, and s_{i2} gives us $d_{i2} = \langle \text{picture, on, the, screen, is, blurry} \rangle$. We then remove "on, "the," and "is" as stopwords, and we obtain the example d_i for e_i as a bag of words $d_i = \langle$LCD, screen, gives, clear, picture, picture, screen, blurry$\rangle$.

3.4.5 Experiments

We tested the proposed method's versatility by conducting experiments on reviews from five different domains: home theater (HT), insurance (I), mattresses (M), cars (C), and vacuums (V). The datasets and feature expressions with their respective groups were obtained from a company that offers opinion mining services. The specifics of each dataset are shown in Table 3.8.

Evaluation Results

In our study, we evaluated our SC-EM approach along with 16 baseline methods. To analyze the impact of having varying numbers of labeled examples (seeds), we conducted experiments by randomly selecting 10%, 20%, 30%, 40%, and 50% of feature expressions from the original dataset as the labeled set L and using the rest as the unlabeled set U. Tables 3.9 and 3.10 present the average results. The best performance in each category is highlighted in bold. Due to limited space, detailed accuracy, purity, and entropy results for only the runs with L comprising 30% of the labeled data (and 70% unlabeled) are presented in Table 3.9. Tables 3.10 summarize the results for all other sizes of the labeled group L, averaged across the five domains. All results were obtained from the test set U, which consisted of the unlabeled data. For the experiments, we used a window size of $t = 5$. In terms of the entropy measure, smaller values indicate better performance, while for purity and accuracy, larger values are better.

The results presented in Tables 3.9 and 3.10 indicate that the proposed algorithm outperformed all 16 baseline methods by a significant margin across all datasets. Specifically, the following observations were made:

Table 3.8 Datasets, feature expressions, and synonym groups

Dataset size	Domains				
	HT*	I*	M*	C*	V*
No. of sentences	6355	12,446	12,107	9731	8785
No. of reviews	587	2802	933	1486	551
No. of feature expressions	237	148	333	317	266
No. of groups (thousands)	15	8	15	16	28

HT* home theater; I* insurance; M* mattresses; C* cars; V* vacuums

Table 3.9 Performance of feature expression classification and clustering methods on various domains (best performance in each category in bold)

	Methods*	Accuracy					Purity					Entropy				
		HT	I	M	C	V	HT	I	M	C	V	HT	I	M	C	V
Unsupervised	SHC	0.20	0.26	0.15	0.17	0.27	0.32	0.41	0.32	0.32	0.38	2.73	2.02	2.69	2.83	2.54
	CHC	**0.26**	**0.35**	**0.26**	**0.26**	**0.40**	0.37	0.46	0.33	0.32	0.43	2.47	2.04	2.56	2.63	2.09
	LDA	0.06	0.11	0.05	0.06	0.03	0.31	0.36	0.32	0.37	0.36	2.54	2.24	2.57	2.39	2.09
	mLSA	0.06	0.14	0.06	0.09	0.03	0.31	0.38	0.34	0.37	0.37	2.53	2.19	2.55	2.40	2.11
	k-means	0.21	0.25	0.15	0.25	0.24	**0.42**	**0.45**	**0.39**	**0.44**	**0.47**	**2.14**	**1.9**	**2.32**	**2.16**	**1.78**
Semisupervised	LDA(L,H)	0.1	0.16	0.10	0.19	0.10	0.32	0.37	0.34	0.39	0.39	2.50	2.22	2.57	2.36	2.09
	DF-LDA(L,H)	0.27	0.25	0.19	0.28	0.31	0.37	0.41	0.39	0.45	0.40	2.32	2.00	2.35	2.15	1.98
	k-means(L, H)	0.2	0.25	0.17	0.27	0.20	0.42	0.43	0.42	0.48	0.48	2.12	1.92	2.26	2.04	1.76
	EM(L, H)	**0.48**	**0.5**	**0.52**	**0.56**	**0.49**	**0.50**	**0.53**	**0.56**	**0.58**	**0.52**	**1.93**	**1.69**	**1.87**	**1.8**	**1.79**
Hard-constrained	Rand(LC, H)	0.42	0.4	0.41	0.36	0.40	0.47	0.47	0.49	0.44	0.57	1.99	1.91	1.94	2.15	1.41
	LDA(LC, H)	0.47	0.43	0.45	0.47	0.43	0.53	0.50	0.52	0.56	0.59	1.81	1.79	1.84	1.76	1.41
	DF-LDA(LC, H)	0.35	0.33	0.23	0.34	0.37	0.49	0.49	0.39	0.51	0.52	1.86	1.71	2.26	1.88	1.58
	k-means(LC, H)	0.52	0.48	0.43	0.45	0.43	0.57	0.55	0.53	0.57	0.63	1.60	1.55	1.79	1.69	1.24
	EM(LC, H)	**0.61**	**0.62**	**0.54**	**0.63**	**0.53**	**0.64**	**0.65**	**0.60**	**0.66**	**0.63**	**1.41**	**1.33**	**1.57**	**1.38**	**1.26**
Soft-constrained	LDA(LC,S)	0.25	0.3	0.26	0.32	0.26	0.35	0.37	0.37	0.39	0.39	2.31	2.04	2.28	2.21	1.86
	k-means(LC,S)	0.33	0.3	0.29	0.29	0.33	0.45	0.43	0.46	0.45	0.53	1.99	1.94	2.10	2.09	1.58
	SC-EM	**0.68**	**0.72**	**0.68**	**0.75**	**0.68**	**0.68**	**0.74**	**0.70**	**0.76**	**0.69**	**1.30**	**1.07**	**1.26**	**1.12**	**1.16**

* See main article for meanings of acronyms and abbreviations

Table 3.10 Performance of feature expression classification and clustering methods with varying percentages of labeled seeds

	Methods	Accuracy					Purity					Entropy				
		10%	20%	30%	40%	50%	10%	20%	30%	40%	50%	10%	20%	30%	40%	50%
Unsupervised	SHC	0.20	0.21	0.21	0.22	0.26	0.33	0.35	0.35	0.37	0.38	2.72	2.63	2.56	2.46	2.25
	CHC	**0.25**	**0.28**	**0.31**	**0.34**	**0.37**	0.35	0.37	0.38	0.40	0.44	2.49	2.39	2.36	2.23	2.08
	LDA	0.07	0.07	0.06	0.06	0.08	0.33	0.33	0.34	0.35	0.38	2.50	2.44	2.37	2.28	2.11
	mLSA	0.07	0.07	0.08	0.07	0.07	0.34	0.35	0.35	0.37	0.38	2.48	2.42	2.36	2.26	2.12
	k-means	0.22	0.23	0.22	0.22	0.22	**0.42**	**0.43**	**0.44**	**0.44**	**0.46**	**2.16**	**2.11**	**2.06**	**1.98**	**1.86**
Semisupervised	LDA(L,H)	0.10	0.10	0.13	0.14	0.15	0.34	0.34	0.36	0.37	0.39	2.48	2.43	2.35	2.25	2.11
	DF-LDA(L,H)	0.23	0.25	0.26	0.27	0.30	0.41	0.40	0.41	0.41	0.44	2.23	2.23	2.16	2.10	1.94
	k-means(L, H)	0.13	0.16	0.22	0.24	0.28	0.42	0.43	0.45	0.45	0.48	2.15	2.11	2.02	1.95	1.79
	EM(L, H)	**0.35**	**0.44**	**0.51**	**0.55**	**0.58**	**0.43**	**0.49**	**0.54**	**0.57**	**0.61**	**2.22**	**1.99**	**1.81**	**1.65**	**1.49**
Hard-constrained	Rand(LC, H)	0.33	0.37	0.40	0.44	0.46	0.40	0.47	0.49	0.54	0.57	2.32	2.00	1.88	1.64	1.50
	LDA(LC, H)	0.36	0.41	0.45	0.49	0.51	0.43	0.50	0.54	0.58	0.61	2.16	1.88	1.72	1.50	1.37
	DF-LDA(LC, H)	0.32	0.33	0.33	0.34	0.36	0.49	0.50	0.48	0.48	0.48	1.90	1.85	1.86	1.83	1.82
	k-means(LC, H)	0.38	0.43	0.46	0.51	0.53	0.44	0.54	0.57	0.61	0.62	1.93	1.69	1.57	1.40	1.32
	EM(LC, H)	**0.45**	**0.53**	**0.58**	**0.63**	**0.64**	**0.51**	**0.60**	**0.64**	**0.69**	**0.70**	**1.92**	**1.53**	**1.39**	**1.17**	**1.08**
Soft-constrained	LDA(LC,S)	0.18	0.23	0.28	0.32	0.36	0.36	0.39	0.37	0.46	0.50	2.37	2.25	2.14	1.90	1.74
	k-means(LC,S)	0.26	0.28	0.31	0.33	0.34	0.44	0.46	0.46	0.48	0.50	2.07	2.01	1.94	1.83	1.69
	SC-EM	**0.49**	**0.62**	**0.7**	**0.74**	**0.81**	**0.54**	**0.65**	**0.71**	**0.76**	**0.82**	**1.90**	**1.49**	**1.18**	**0.99**	**0.73**

- The experiments revealed that the performance of SHC, CHC, LDA, mLSA, and k-means without seeds (i.e., no labeled data) was the worst. As expected, using seeds improved the results. Without seeds, DF-LDA and LDA performed similarly.
- Our findings suggest that LDA-based methods performed the weakest among the tested methods. While k-means methods showed slight improvement, EM methods were the best performers. This indicates that classification (EM) outperforms clustering. DF-LDA's and k-means' results were found to be similar.
- In the case of LDA and k-means, hard-constrained methods (LDA(LC,H) and K-means(LC,H)) outperformed soft-constrained methods (LDA(LC,S) and K-means(LC,S)). This suggests that soft constraints may lead to some correctly constrained expressions being assigned to the wrong groups. However, for EM methods, the soft-constrained method (SC-EM) performed significantly better than the hard-constrained version (EM(LC,H)). This indicates that the Bayesian classifier used in EM can leverage the soft constraints to correct some incorrect assignments.
- The comparatively weaker performance of Rand(LC,H) as compared to SC-EM in various settings indicates that constraints alone (synonyms and word sharing) are insufficient in improving the results. Our findings suggest that EM can considerably enhance the results.
- Our analysis of the EM-based methods revealed that the use of soft seeds in SL had a significant impact on all datasets. In particular, SC-EM was found to be the best-performing method.
- Our experiments revealed that increasing the number of labeled examples from 10% to 50% led to an improvement in the performance of all methods except for DF-LDA, which showed minimal improvement.

We conducted further experiments by varying the text window size t from 1 to 10 to assess its impact on the performance of SC-EM. The results are presented in Fig. 3.11, and it is evident that window sizes between 2 to 5 produced similarly good results. Therefore, we used $t = 5$ for all our evaluations.

3.4.6 Conclusion

Our proposed method was evaluated empirically using five real-life datasets, and the results demonstrated its superiority over 16 baseline methods. In our future research, we aim to enhance the accuracy of our approach by leveraging additional natural-language knowledge at the semantic level.

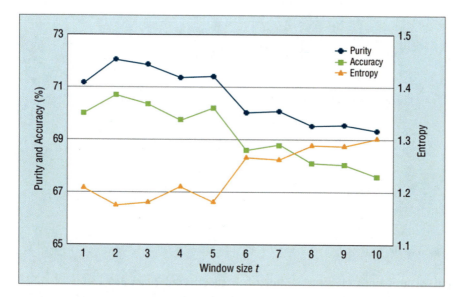

Fig. 3.11 Influence of the context window size on SC-EM

3.5   Exploiting Effective Features for Chinese Sentiment Classification

3.5.1   Introduction

The need to handle and analyze opinionated texts on the internet has led to the development of sentiment analysis, also known as opinion mining, which has become a crucial technology [24]. However, an outstanding issue in machine learning-based sentiment classification is how to extract complex features that outperform simple ones, and determining which types of features are more valuable [25]. Although the majority of existing research focuses on simple features such as single words [26], character N-grams [27, 28], word N-grams [25, 27, 29, 30], or a combination of these features, few studies systematically compare and statistically analyze the impact of these different types of features. This section presents an in-depth study of all types of features to identify effective features for sentiment classification and address these questions through experimental analysis.

3.5.2   Methodology

Figure 3.12 presents the framework of sentiment classification based on supervised learning, which comprises two phases: training and classifying. During the training phase, a sentiment classifier is developed using labeled documents. In the classifying

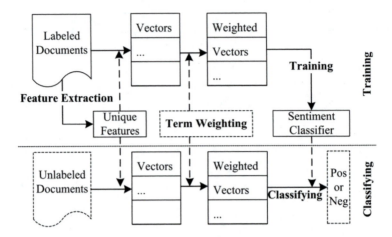

Fig. 3.12 The framework of sentiment classification based on supervised learning

phase, the trained classifier is utilized to classify unlabeled documents as either positive or negative. We provide a detailed explanation of the essential stages of the framework, which encompass feature extraction, term weighting, training, and classification.

Feature Extraction

The objective of this stage is to derive distinctive characteristics from the annotated records. As illustrated in Fig. 3.12, these particular attributes are utilized to convert the initial documents into vectors, which serve as the foundation for both the training and classification processes. Consequently, this phase plays a critical role in determining the effectiveness of the sentiment classification.

In this section, in addition to the frequently employed N-gram features, the sentiment term, substring, substring-group, and key-substring-group features extraction algorithms are also examined.

1. **Sentiment term features**: Sentiment classification heavily relies on sentiment words and phrases, as they are the primary indicators of sentiment [31]. Therefore, it is reasonable to use pre-existing sentiment lexicons as features for sentiment classification. To be more precise, the extraction process involves the following steps. The initial step involves extracting all terms from the training corpus to create a pool of potential features. Next, the extracted terms undergo a selection process. For each term, it is checked whether it appears in the sentiment lexicon. If a term is present in the lexicon, it is chosen as a feature, while those that are not are disregarded. In this study, a trustworthy sentiment lexicon from Tsinghua University is employed, which comprises 5563 positive words and 4464 negative words. This technique is referred to as the baseline method in the subsequent experiments.

2. **N-gram features**: N-gram features are frequently utilized in text classification tasks [26–30]. The N-gram extraction algorithm is straightforward, and Algorithm 3.9 displays the pseudocode for this algorithm. In this algorithm, $S[j]$ denotes the *j*th unit in the string S, and n represents the N-gram's size.

Algorithm 3.9: N-Gram Features Extraction Algorithm

Input: $S[1, \cdots, l]$
Output: $F[1, \cdots, p], p = l - n + 1$
1:    **for** $i = 1$ to $l - n + 1$ **do**
2:        F.insert($S[i, \cdots, i - n + 1]$)
3:    **end for**
4:    **return** F

N-gram features can be classified into character N-gram features and word N-gram features, depending on the level of granularity. For instance, consider the sentence "I like this camera." The character bigram features of this sentence are Il, li, ik, ke, et, th, hi, is, sc, ca, am, me, er, ra, while its word bigram features are "I like," "like this," and "this camera." In experiments, character-unigram (CU), character-bigram (CB), character-trigram (CT), word-unigram (WU), word-bigram (WB), and word-trigram (WT) features are all investigated.

3. **Substring Features**: Although substring features are not frequently utilized in text classification tasks, they have numerous potential advantages [32]. Algorithm 3.10 displays the pseudocode for the substring feature extraction algorithm. In this algorithm, $S[j]$ denotes the *j*th unit in the string S. It is noteworthy that the substring features comprise all the possible combinations of n-gram features, where $n - \text{gram}$ from 1 to l

Algorithm 3.10: Substring Features Extraction Algorithm

Input: $S[1, \cdots, l]$
Output: $F[1, \cdots, p], p = \frac{l(l+1)}{2}$
1:    **for** $n = 1$ to l
2:        **for** $i = 1$ to $l - n + 1$ **do**
3:            F.insert($S[i, \cdots, i - n + 1]$)
4:        **end for**
5:    **end for**
6:    **return** F

In the aforementioned sentence, the word substring features are "I," "I like," "I like this," "I like this camera," "like," "like this," "like this camera," "this," "this camera," and "camera." The strings within a document can be tokenized into either characters or words, resulting in two types of features: character-based substring features (CS) and word-based substring features (WS), respectively.

4. **Substring-group features**: A corpus D with a length of l contains $\frac{l(l+1)}{2}$ possible substrings. When l is sizable, the number of substring features can become excessively large for classification purposes. However, suffix tree techniques can be used to cluster substring features into a relatively small number of

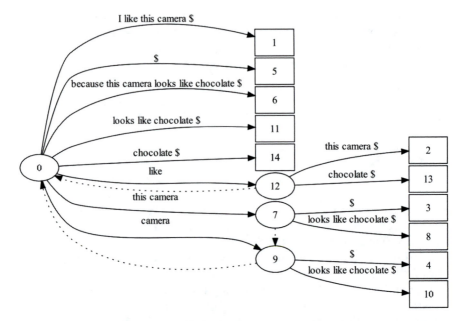

Fig. 3.13 Generalized suffix tree constructed by the two short sentences: "I like this camera" and "because this camera looks like chocolate." The dotted arrows represent suffix links

equivalence groups. The primary steps involved in this process are outlined below.

To begin, Ukkonen's algorithm with $O(l)$ time complexity is used to construct a generalized suffix tree that encompasses all the documents in the training corpus [33]. In this section, a short sentence serves as the fundamental unit of sequence added to construct the suffix tree. This sentence is obtained by splitting longer sentences using punctuation marks such as periods, commas, question marks, exclamation marks, ellipses, semicolons, and colons.

Next, all the substrings from the training and testing documents are compared to the constructed suffix tree, and the IDs of all the matched nodes are recorded as features for the corresponding document. By leveraging the suffix links to move from one suffix to the next suffix, this process can also be completed in $O(l)$ time.

As stated in [34], for a text corpus with a length of l, the constructed suffix tree comprises l leaf-nodes and at most $l - 1$ internal nodes. Therefore, the number of features can be reduced from $\frac{l(l+1)}{2}$ to $(2l - 1)$ through the use of suffix tree techniques.

Consider the sentence "I like this camera, because this camera looks like chocolate" as an example. The suffix tree created using the two short sentences is depicted in Fig. 3.13, resulting in 14 substring groups. When the sentence "I like chocolate very much" is matched against this tree, nodes 1, 12, 13, and 14 are identified. Consequently, the substring-group features for this sentence are "1, 12, 13, 14."

Table 3.11 The criteria for key-node selection

L	The minimum frequency. A node is not selected, if it has less than L leaf-nodes in the suffix tree
H	The maximum frequency. A node is not selected, if it has more than H leaf-nodes in the suffix tree
B	The minimum number of children. A node is not selected, if it has less than B children
P	The maximum parent–child conditional probability. A node u is not selected, if the probability $Pr((v|u)) = freq(v)/freq(u) \geq P$, where u is the parent node of v
Q	The maximum suffix-link conditional probability. A node $s(v)$ is not selected, if the probability $Pr((v|s(v))) = freq(v)/freq(s(v)) \geq Q$, where the suffix-link of v points to $s(v)$

The aforementioned short sentence can be tokenized into either characters or words. If the suffix tree is built using the character sequence of short sentences, the extracted features are referred to as character-based substring-group features (CSG). Conversely, if the suffix tree is built using word sequences, the extracted features are known as word-based substring-group features (WSG).

5. **Key-substring-group features**: While $(2l - 1)$ is a significant reduction from the original $\frac{l(l+1)}{2}$ substring features, it can still be impractical for classification when the training corpus is sizable. It is evident that the nodes in the suffix tree are still excessive. To address this issue, the criteria suggested in [32] are employed to further decrease the number of substring groups.

Compared to the substring-group extraction algorithm, the key-substring-group extraction algorithm incorporates an extra step at the end that eliminates certain IDs based on the criteria listed in Table 3.11. The resultant removed IDs are then utilized as the key substring features for the corresponding document.

In experiments, the suggested parameter values for the criteria are utilized: $L = 20$, $H = 8000$, $B = 8$, $P = 0.8$ and $Q = 0.8$. Additionally, as previously mentioned in the Substring-group features subsection, the extracted features can be classified into either character-based key-substring-group features (CKSG) or word-based key-substring-group features (WKSG).

Term Weighting

The representation of a problem can significantly influence the overall accuracy of learning systems [35]. In the case of documents, which are typically presented as strings of characters, it is necessary to convert them into a suitable representation that is compatible with learning algorithms and classification tasks.

A widely adopted technique in text classification is tfidf $- c$, which is a variant of the standard tfidf method.

$$\text{tfidf} - c : \frac{tf\left(t, d_j\right), \log \frac{N}{df(t)}}{\sqrt{\sum_{t \in d_j} \left(tf\left(t, d_j\right), \log \frac{N}{df(t)}\right)^2}} \tag{3.33}$$

In tfidf $-$ c, the notation t_k represents a unique term corresponding to a single feature, while $tf(t_k, d_j)$ represents the frequency with which term t_k appears in the document d_j. Additionally, $df(t_k)$ denotes the number of documents in which the term t_k appears, and N represents the total number of training documents.

Training and Classifying

During this stage, machine learning algorithms are utilized to train the sentiment classifier, which can subsequently predict the classes of unlabeled test documents.

In comparison to other state-of-the-art methods such as Naive Bayes, Maximum Entropy, K-Nearest Neighbor, and Neural Network, SVMs have demonstrated significant performance improvements [26, 27, 29, 30]. Due to their robustness, SVMs have obviated the need for costly parameter tuning [36]. Additionally, SVMs employ an overfitting protection technique that is independent of the number of features, allowing them to potentially handle vast feature spaces.

Given the impressive performance of SVMs, they are utilized in this section for sentiment classification. Training and testing of the SVM model are conducted using the SVM[light] package, with default parameters.

3.5.3 Experimental Setup

Data Sets

The experiments are carried out on two authoritative datasets from two distinct domains: Hotel and Product.

The initial dataset is obtained by crawling Ctrip, one of the most prominent Chinese websites for hotel and flight bookings, while the second dataset is collected from IT168, a popular Chinese IT product website. To classify the reviews as either positive or negative, both datasets undergo a manual filtering process. A summary of each dataset is presented in Table 3.12.

Table 3.12 The summary of the data sets

Dataset name	#Positive	#Negative	Publish authority
Hotel	2000	2000	Chinese academy of science
Product	451	435	Peking university

For the experiments, each dataset is randomly split into three equal-sized folds, ensuring that the class distribution is balanced in each fold.

Evaluation Metrics

To evaluate the performance of the sentiment classification, the standard precision, recall, and F1 scores are utilized to measure the performance of the negative and positive classes, respectively. Additionally, the overall performance of the sentiment classification is evaluated using accuracy. These metrics are equivalent to those employed in general text categorization.

3.5.4 Experimental Results

The experiments conducted in this study closely analyze a variety of features to determine the most effective ones for sentiment classification. Specifically, the baseline, CU, CB, CT, WU, WB, WT, CS, WS, CSG, WSG, CKSG, and WKSG features are all carefully examined.

The features employed in this study can be broadly classified into two main categories: N-gram-based features and substring-based features. Initially, this section analyzes the performances of these two feature categories, after which representative features from each category are selected for comparison at the end of the section.

Performances of N-Gram-Based Features

Table 3.13 presents the performance of the N-gram-based features, with CB consistently outperforming the others in terms of accuracy, precision, recall, and F1. WU is capable of achieving acceptable accuracy while utilizing a relatively small number of features, whereas WB consistently underperforms CB. Tri-gram features (CT and WT) exhibit the poorest performance, both in terms of performance and number of features. Considering both performance and number of features, CB and WU are selected for further comparisons.

The results also suggest that the numbers of CB (or CT) are comparable to those of WB (or WT), but the performance of WB (or WT) is consistently inferior to that of CB (or CT).

Performances of Substring-Based Features

Table 3.14 presents the performances of the substring-based features. From the data, it is evident that the substring group features (CSG and WSG) outperform all other substring-based features. When compared to CS (or WS), CSG (or WSG) has an

Table 3.13 Performance of the *N*-gram-based features on two data sets

	Method	Positive			Negative			Total	
		P	R	F1	P	R	F1	accuracy	#Features
Hotel	Baseline	0.857	0.810	0.833	0.820	0.865	0.842	0.838	1345
	CU	0.882	0.866	0.874	0.868	0.884	0.876	0.875	3216
	CB	0.927	0.910	0.918	0.912	0.928	0.920	0.919	68,279
	CT	0.926	0.899	0.912	0.902	0.928	0.915	0.913	172,495
	WU	0.905	0.893	0.899	0.894	0.906	0.900	0.899	10,293
	WB	0.928	0.894	0.911	0.898	0.930	0.914	0.912	86,744
	WT	0.886	0.867	0.876	0.870	0.888	0.879	0.878	146,213
Product	Baseline	0.849	0.763	0.804	0.777	0.858	0.815	0.810	475
	CU	0.920	0.903	0.911	0.902	0.917	0.909	0.910	1627
	CB	0.932	0.914	0.923	0.913	0.930	0.921	0.922	17,713
	CT	0.864	0.876	0.869	0.870	0.854	0.861	0.865	30,739
	WU	0.938	0.920	0.929	0.919	0.936	0.927	0.928	3392
	WB	0.863	0.882	0.872	0.876	0.853	0.864	0.868	15,730
	WT	0.675	0.883	0.765	0.821	0.556	0.662	0.723	19,884

Table 3.14 Performance of the substring-based features on two data sets

	Method	Positive			Negative			Total	
		P	R	F1	P	R	F1	accuracy	#Features
Hotel	Baseline	0.857	0.810	0.833	0.820	0.865	0.842	0.838	1345
	CS	0.882	0.866	0.874	0.868	0.884	0.876	0.875	3216
	WS	0.927	0.910	0.918	0.912	0.928	0.920	0.919	68,279
	CSG	0.926	0.899	0.912	0.902	0.928	0.915	0.913	172,495
	WSG	0.905	0.893	0.899	0.894	0.906	0.900	0.899	10,293
	CKSG	0.928	0.894	0.911	0.898	0.930	0.914	0.912	86,744
	WKSG	0.886	0.867	0.876	0.870	0.888	0.879	0.878	146,213
Product	Baseline	0.849	0.763	0.804	0.777	0.858	0.815	0.810	475
	CS	0.920	0.903	0.911	0.902	0.917	0.909	0.910	1627
	WS	0.932	0.914	0.923	0.913	0.930	0.921	0.922	17,713
	CSG	0.864	0.876	0.869	0.870	0.854	0.861	0.865	30,739
	WSG	0.938	0.920	0.929	0.919	0.936	0.927	0.928	3392
	CKSG	0.863	0.882	0.872	0.876	0.853	0.864	0.868	15,730
	WKSG	0.675	0.883	0.765	0.821	0.556	0.662	0.723	19,884

absolute advantage in the number of features utilized for classification, highlighting the effectiveness of the suffix tree technique in reducing substring features.

Another noteworthy finding is that word-based substring features (WS, WSG) generally exhibit superior performance to character-based substring features (CS, CSG), in terms of accuracy, precision, recall, F1, and the number of features (#Features). This performance discrepancy is a direct reflection of the influence of Chinese word segmentation techniques on classifiers that utilize these features.

Additionally, the key substring features (CKSG, WKSG) exhibit relatively strong performance despite utilizing only a small number of features.

Following the aforementioned analysis, WSG, CKSG, and WKSG are selected for comparison purposes.

Comparison

To ascertain the effectiveness of different feature types, this subsection conducts a comprehensive comparison between each category's representations and the baseline. The comparison results are presented in Figs. 3.14 and 3.15. The trend that precision, recall, and F1 scores follow is consistent with that of accuracy, as illustrated by the data in Tables 3.13 and 3.14.

Based on their respective performances and number of features, these six features can be further classified into two groups: a high-performance group (CB, WU, WSG) and a low-cost group (baseline, WKSG, CKSG).

Regarding the high-performance group, WSG has a smaller number of features compared to CB and WU. Nevertheless, WSG achieves significantly better accuracy (2%) than CB on the Product dataset, while obtaining nearly identical performance

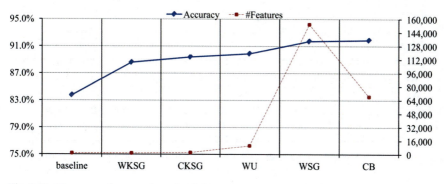

Fig. 3.14 The comparison results on hotel data set

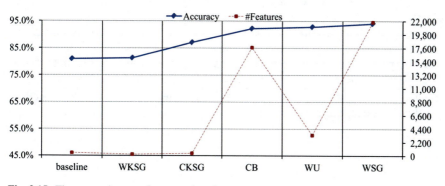

Fig. 3.15 The comparison results on product data set

on the Hotel dataset. Furthermore, WSG outperforms WU on both the Product and Hotel datasets. These findings underscore the potential of substring features in sentiment classification.

In the low-cost group, CKSG features exhibit superior performance to both the baseline and WKCG features on both datasets. These results provide compelling evidence of the efficacy of character key-substring-group features.

The sentiment term-based features (baseline) exhibit the weakest performance in both the high-performance and low-cost groups. Specifically, the baseline features result in the lowest accuracy score, and they are surpassed by WKSG features in terms of the number of features utilized. This outcome is contrary to our initial expectations and provides an answer to the third question raised in the introduction section.

3.5.5   Conclusion

This section concludes an evaluation of various feature types utilized for sentiment classification. In our research, we explore sentiment word-based, substring-based, substring-group-based, and key-substring-group-based features, which have not been extensively studied in the context of sentiment classification, in addition to the commonly used N-gram features.

Experiments were conducted on two authoritative datasets from different domains to evaluate the effectiveness of these features in Chinese sentiment classification. The experimental results revealed that different types of features possess distinct discriminatory capabilities in sentiment classification. Specifically, the following conclusions can be drawn from analyzing the results of the experiments: (1) Character bigram features (CB) consistently outperform other N-gram features. (2) While character bigram features (CB) demonstrate promising performance, substring-group features have greater potential to improve sentiment classification performance by combining substrings of varying lengths. (3) Sentiment words and phrases extracted from existing sentiment lexicons are not effective for sentiment classification. (4) Features of varying lengths are typically more effective than those of fixed lengths.

In addition to these critical findings, this section introduces a set of substring-group-based features for sentiment classification that have not been explored previously. The efficacy of these features is thoroughly investigated, shedding light on the underlying mechanisms of their effectiveness. Furthermore, this section provides a comprehensive overview of all the feasible techniques for sentiment classification based on machine learning.

3.6 An Empirical Study of Unsupervised Sentiment Classification of Chinese Reviews

3.6.1 Introduction

Thus far, the majority of sentiment classification techniques are based on supervised learning, owing to its superior performance compared to unsupervised approaches. However, unsupervised learning methods offer several unique advantages that supervised methods lack. For instance, unsupervised techniques can classify opinionated texts at the word or sentence level, whereas supervised approaches may not be suitable due to a lack of training corpora at these levels. Additionally, unsupervised approaches can be utilized for feature-based opinion mining, whereas supervised methods tend to be less effective in this regard [37].

The primary objective of this section is to conduct an empirical investigation of unsupervised sentiment classification of Chinese reviews and enhance the classification performance utilizing limited available sentiment resources in Chinese. To achieve this goal, the study evaluates all existing Chinese sentiment lexicons, both individually and in combination, using the proposed framework. Additionally, domain-specific sentiment noise words are identified and eliminated using unlabeled data to improve classification performance. Experiments are conducted on three open datasets from two domains, and the results reveal that the proposed algorithm for sentiment noise word removal can substantially enhance classification performance.

3.6.2 Proposed Technique

Algorithm 3.11: The Framework of Lexicon-Based Unsupervised Sentiment Classification in Chinese

---

Input: Test corpus T; Unlabeled corpus U; Positive lexicon and weight Pos, posWeight; Negative lexicon and weight Neg, negWeight; Negation lexicon Negation; Intensifier lexicon and weight Intensifier ρ

Output: Sentiment polarities of the reviews in T

Step 1 SNW identification

(1)   Construct a graph: The nodes are all the words in both Pos and Neg. The edges are the co-occurrence relationship extracted from the unlabeled corpus U.

(2)   Run graph-ranking based algorithm to calculate the rank value for each nodes.

(3)   Select the top one per mil nodes as sentiment noise words set SNW.

Step 2 Sentiment polarity computation for T

        For each review rev_k in the test corpus T:

(1)   Tokenize rev_k into short sentence set S, and split each sentence $s \in S$ into word set W_s.

(2)   For any word $w \notin SNW$ in a sentence $s \in S$, compute its sentiment polarity $sp(w)$ as follows:

        (a) If $w \notin Pos$ & $w \notin Neg$, $sp(w) = 0$;

(continued)

If $w \in Pos$ & $w \notin Neg$, $sp(w) = $ posWeight ;
If $w \notin Pos$ & $w \in Neg$, $sp(w) = $ negWeight; Otherwise,
$sp(w) = $ posWeight + negWeight .

(b) Within the window of q words revious to w, if there is a term $w' \in Neg$,
$sp(w) = -sp(w)$.

(c) Within the window of q words pervious and next to w, if there is a term $w' \in$ Intensifier, $sp(w) = \rho \times sp(w)$.

(3) Compute rev_k sentiment polarity:

$$sp(rev_k) = \begin{cases} \text{positive, } \sum_{s \in S} \sum_{w \in W_s} p(w) > 0 \\ \text{negative, } \sum_{s \in S} \sum_{w \in W_s} p(w) < 0 \\ \text{random, otherwise} \end{cases}$$

Algorithm 3.11 illustrates the framework of the proposed technique. The input consists of a test corpus, an unlabeled corpus, and lexicons that encompass positive lexicons, negative lexicons, negation lexicons, and intensifier lexicons. The output is the sentiment polarities of the test corpus. The algorithm comprises two primary steps. First, the sentiment noise words (SNWs) are identified using the unlabeled corpus U. Second, the sentiment polarity of each review is calculated based on the identified SNWs and the input lexicons.

SNW Identification

To address the domain dependence issue of sentiment noise words (SNWs), natural language knowledge is leveraged to mitigate this problem. Specifically, frequent co-occurrences of sentiment words with other words in short sentences are likely to be indicative of SNWs. Drawing on this empirical knowledge, domain-specific unlabeled corpora are utilized to identify the SNWs. The identification process is outlined below.

Initially, an undirected graph G is created to represent the relationships among all the sentiment words. The nodes comprise all the sentiment words found in both the positive and negative lexicons, while the edges correspond to co-occurrence relationships extracted from the unlabeled corpus. The weight of an edge (w_i, w_j) is determined by the number of co-occurrences of w_i and w_j. Specifically, all reviews in the unlabeled corpus are tokenized into short sentences. If two sentiment words are present in the same sentence, the weight of the edge between these two words is incremented by one unit. An adjacency matrix M is employed to describe G, and $M = (M_{i, j})_{n \times n}$ is defined as:

$$M_{ij} = \begin{cases} \text{Weight}(w_i, w_j), i \neq j; \\ 0, \text{otherwise} \end{cases} \tag{3.34}$$

Afterwards, M is normalized to $\tilde{M}$ to ensure that the sum of each row equals 12–1:

$$\tilde{M}_{i,j} = \begin{cases} M_{i,j}/\sum_{j=1}^{n} M_{i,j}, \sum_{j=1}^{n} M_{i,j} \neq 0; \\ 0, \text{otherwise} \end{cases} \qquad (3.35)$$

Next, the graph-ranking algorithm is executed, which generates a score value for each sentiment word. The algorithm can be represented in a recursive form:

$$\text{Score}_{\text{all}}(w_i) = d \cdot \sum_{j \neq i} \text{score}_{\text{all}}(w_i) \cdot \tilde{M}_{j,i} + \frac{1-d}{n} \qquad (3.36)$$

And the matrix form is

$$\lambda = \lambda \tilde{M}^T \lambda + \frac{1-d}{n} e \qquad (3.37)$$

In the above equation, $\lambda = \text{score}_{\text{all}}(w_i)_{n \times 1}$ represents the vector of scores, while e denotes a unit vector with all elements equal to $1/n$. The damping factor is represented by d. The initial scores of all sentiment words are set to 1, and the iterative algorithm in Eq. (3.36) is utilized to compute new scores for all nodes in G. The damping factor is fixed at 0.85, and the number of iterations is set to 10,000 for implementation purposes.

Lastly, due to the potent discriminating value of adjective sentiment words in sentiment analysis, the top 1/1000 nonadjective words are identified as noise words based on their scores.

Sentiment Polarity Computation

The proposed framework involves a second main step where sentiment weights are assigned to positive and negative words, referred to as posWeight and negWeight respectively. Through empirical observation, it was found that negative words tend to have a greater impact on the overall sentiment polarity of reviews than positive words. As a result, posWeight was set to 1 while negWeight was set to -2. The intensifier terms ρ were given a weight of 2. The window size q, which determines the range of influence of negation and intensifier terms on sentiment words, was set to 2.

Algorithm 3.11's Step 2 utilizes the SNW generated in Step 1 along with the parameters specified earlier to calculate the sentiment polarity of reviews based on input lexicons. These lexicons, which are crucial in determining the algorithm's output, can be classified into three categories: negation lexicon, intensifier lexicon, and sentiment lexicon. The Chinese negation lexicon consists of 13 negation terms sourced from prior research, while the Chinese intensifier lexicon comprises 148 terms collected from the HOWNET's Chinese Vocabulary for Sentiment Analysis.

Table 3.15 The summary of all sentiment lexicons

Lexicon name	Number		
	Positive words	Negative words	Total words
D1	836	1254	2081
D2	3730	3116	6770
D3	5563	4464	10,027
D1 + D2	4528	4320	8746
D1 + D3	6144	5434	11,521
D2 + D3	8001	6709	14,581
D1 + D2 + D3	8560	7642	16,000

Table 3.16 Summary of the datasets

Dataset name	Number	
	Labeled corpus	Unlabeled corpus
Product reviews	886 (dataset1)	1000 (dataset1)
Hotel reviews	2000 (dataset2)	16,000(dataset3)

There are three publicly available Chinese sentiment lexicons. Among these, two were developed by HOWNET, while the third was created by Tsinghua University. Table 3.15 provides an overview of these lexicons and their possible combinations. In cases where there are overlapping words, they are eliminated from the combined lexicon.

3.6.3 Empirical Evaluation

This section focuses on investigating the impact of sentiment noise words and sentiment lexicon size on the proposed framework through experiments conducted on three datasets obtained from two different domains.

Datasets

The proposed framework is evaluated through experiments conducted on three publicly available datasets from two distinct domains. Specifically, one dataset pertains to product reviews while the other two datasets involve hotel reviews. Table 3.16 provides a brief overview of these datasets, with labeled corpora being utilized as test sets and unlabeled corpora being utilized for sentiment noise word identification.

The product review dataset consists of both labeled and unlabeled corpora, while the two hotel review datasets are solely labeled. However, an unlabeled corpus is necessary to identify sentiment noise words. For the purpose of the experiments, one of the hotel review datasets is utilized as an unlabeled corpus, with its class label information being disregarded. As these two hotel review datasets were obtained from different institutions and released at different times, they are considered to be independent and can be combined into a single group for analysis.

Evaluation Measures

As sentiment classification is a form of text categorization, the overall performance of the presented algorithms is evaluated using macro precision, macro recall, macro F1, and accuracy metrics. These evaluation metrics are similar to those used in general text categorization and can be found in greater detail in [37].

Impact of the SNW

The impact of sentiment noise words (SNWs) on unsupervised sentiment classification is analyzed in the first set of experiments. The proposed framework in Algorithm 3.11 is utilized under two different settings: original lexicon (OL) and filtered lexicon (FL). In the OL setting, SNWs are not taken into account, and Step 1 in Algorithm 3.11 is skipped, with SNW set to null in Step 2. In contrast, in the FL setting, SNWs are identified using the corresponding unlabeled data in Step 1 and filtered in Step 2.

Table 3.17 presents the performance of the OL setting, while Table 3.18 displays the difference in performance between the FL and OL settings. The proposed algorithm in Algorithm 3.11 involves random values in substep 3 of Step 2, and

Table 3.17 Experimental results based on the original sentiment lexicons OL (without SNW removal)

		Macro precision	Macro recall	Macro F1	Accuracy
Product	D1	0.62	0.61	0.61	0.62
	D2	0.79	0.76	0.76	0.77
	D3	0.80	0.79	0.79	0.79
	D1 + D2	0.79	0.77	0.77	0.77
	D1 + D3	0.81	0.80	0.80	0.80
	D2 + D3	0.82	0.80	0.79	0.80
	D1 + D2 + D3	0.82	0.81	0.81	0.81
Hotel	D1	0.65	0.65	0.65	0.65
	D2	0.76	0.74	0.74	0.74
	D3	0.76	0.76	0.76	0.76
	D1 + D2	0.77	0.76	0.76	0.76
	D1 + D3	0.77	0.77	0.77	0.77
	D2 + D3	0.74	0.74	0.74	0.74
	D1 + D2 + D3	0.77	0.76	0.76	0.76

Table 3.18 Improvements made by SNW removal

		Macro precision	Macro recall	Macro F1	Accuracy
Product	D1	0.01	0.01	0.01	0.01
	D2	0.03	0.03	0.03	0.03
	D3	0.02	0.02	0.02	0.02
	D1 + D2	0.02	0.03	0.04	0.04
	D1 + D3	0.01	0.01	0.01	0.01
	D2 + D3	0.03	0.04	0.04	0.04
	D1 + D2 + D3	0.01	0.02	0.02	0.02
Hotel	D1	0.02	0.02	0.02	0.02
	D2	0.02	0.04	0.04	0.04
	D3	0.02	0.02	0.02	0.02
	D1 + D2	0.02	0.03	0.03	0.03
	D1 + D3	0.01	0.01	0.01	0.01
	D2 + D3	0.03	0.04	0.04	0.04
	D1 + D2 + D3	0.03	0.03	0.03	0.03

therefore, the results reported in this section are the average values obtained over 10 runs.

The results in Table 3.18 indicate that all values are positive, ranging from 1% to 4%. This suggests that the FL setting outperforms the OL setting, implying that SNWs have a deleterious effect on unsupervised sentiment classification. Furthermore, the proposed algorithm for removing SNW is effective in improving classification performance (Step 1 in Algorithm 3.11), which is an intuitive observation. Additionally, since unlabeled data are used in eliminating SNW, the enhanced performance highlights the potential of unlabeled data to enhance unsupervised classification accuracy.

It is worth noting that even a slight improvement in percentage is noteworthy, particularly since we are approaching the upper limit of accuracy. This finding underscores the importance of existing sentiment lexicons' accuracy in sentiment classification, and highlights the need to identify and eliminate noise words from these lexicons.

Domain-Dependent Characteristics of SNW

This subsection explores another important characteristic of SNW, namely, domain dependence. Experiments are conducted on both product reviews and hotel reviews, with three different settings: OL, FL, and FL.' FL is similar to FL in that SNW are identified in Step 1 and filtered in Step 2 of Algorithm 3.11. However, the identification of SNW differs between FL and FL.' In substep 1 of Step 1, FL uses the corresponding unlabeled data, while FL' uses noncorresponding unlabeled data. In other words, SNW identified by FL are from the same domain, while SNW identified by FL' are from different domains. For example, in experiments on product reviews,

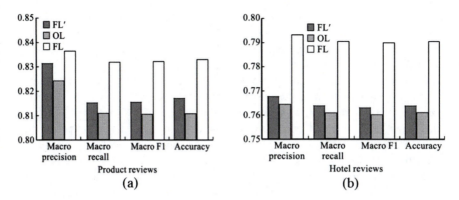

Fig. 3.16 Domain-dependent characteristics of SNW

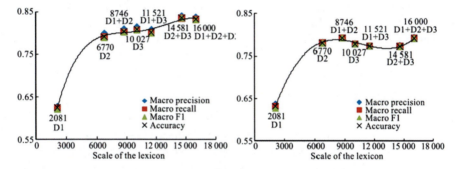

Fig. 3.17 Influence of the lexicons' scale

SNW in FL are extracted using an unlabeled corpus from the same product domain, while SNW in FL' are extracted using an unlabeled corpus from the hotel domain.

As the trends observed in other individuals and combinations (Table 3.15) are similar to those in D1 + D2 + D3, only the results produced by this combination are shown in Fig. 3.16. The following observations can be made from Fig. 3.16: (1) Both FL and FL' perform significantly better than OL, indicating that SNW always interfere with unsupervised sentiment classification. (2) FL outperforms FL' by a large margin, even though they both consider SNW. This finding suggests that SNW are domain-dependent, as the only difference between FL and FL' is the domain of their SNW.

Influences of the Sentiment Lexicons' Scale

The influence of sentiment lexicon size on unsupervised sentiment classification is investigated in this section. Experiments are carried out on product review and hotel review datasets. Figure 3.17 displays the performance of each individual and combined sentiment lexicon in Table 3.15 on both product and hotel reviews. As the FL setting performs the best, it is adopted for all subsequent experiments.

Fig. 3.17 indicates a clear relationship between sentiment lexicon size and classification performance. As the scale of sentiment lexicons increases, the performance of sentiment classification improves significantly, particularly in experiments conducted on the product reviews dataset. Additionally, the following observations can be made: (1) Among individual lexicons D1, D2, and D3, D3 performs the best, while D1 performs the worst, in terms of all evaluation metrics in every experiment. (2) Among lexicon combinations, D1 + D2 + D3 consistently outperforms other combinations. (3) Combined lexicons outperform individual lexicons in all cases, as demonstrated in Fig. 3.17. (4) Interestingly, D1 performs poorly in classification by itself, but when combined with other lexicons, it has a synergistic effect. For instance, D1 + D2 performs better than D2 alone, and D1 + D2 + D3 outperforms D2 + D3.

Based on these observations of individual lexicons and their combinations, it can be concluded that sentiment classification performance is positively correlated with Chinese sentiment lexicon size.

3.6.4 Conclusion

This section presents a framework for lexicon-based unsupervised sentiment classification of Chinese reviews. An empirical study is conducted to examine the impact of sentiment lexicon accuracy and scale on classification. To obtain accurate sentiment lexicons for classification, domain sentiment noise words are filtered out using natural language knowledge. Experiments are conducted on two domains, and the results indicate that the proposed algorithm for filtering sentiment noise words enhances classification performance. This finding underscores the importance of removing noise words from existing sentiment lexicons. Additionally, the experimental results highlight the domain dependence of sentiment noise words.

Furthermore, all relevant Chinese lexicons are collected and evaluated under the proposed framework. The experimental outcomes demonstrate that sentiment classification performance is positively correlated with sentiment lexicon scale, and that combinations of sentiment lexicons typically yield better performance than individual lexicons.

In future work, we plan to expand the seed sentiment lexicon and leverage unlabeled corpora to enhance sentiment classification performance. Specifically, we will explore methods for automatically identifying and incorporating new sentiment words into the lexicon using unsupervised learning techniques. Additionally, we will investigate the use of semisupervised learning methods in conjunction with unlabeled data to improve sentiment classification accuracy. Overall, our goal is to further advance the field of unsupervised sentiment classification in Chinese, and to enable more accurate and efficient analysis of large-scale Chinese text data.

3.7   Feature Subsumption for Sentiment Classification in Multiple Languages

3.7.1   Introduction

In the realm of machine learning-based sentiment classification, previous studies have primarily focused on character or word N-grams features, with little attention given to substring-group features. This study introduces the use of substring-group features for sentiment classification, achieved through a transductive learning-based algorithm for feature extraction and selection. To demonstrate the algorithm's generality, experiments are conducted on three open datasets in three different languages: Chinese, English, and Spanish. The experimental results show that the proposed algorithm outperforms the best performance in related work, and the proposed feature subsumption algorithm for sentiment classification is multilingual. Additionally, the results indicate that the transductive learning-based algorithm can significantly improve sentiment classification performance compared to the inductive learning-based algorithm. With regards to term weighting, the experiments demonstrate that the "tfidf-c" approach outperforms all other term weighting methods in the proposed algorithm.

3.7.2   The Proposed Algorithm

The proposed algorithm comprises four stages, as depicted in Fig. 3.18. These stages include substring-group feature extraction, term weighting, feature selection, and classification.

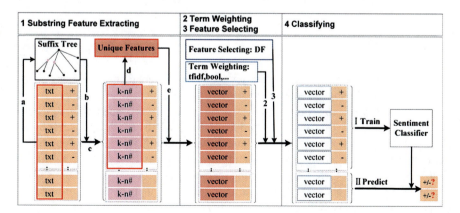

Fig. 3.18 The framework of the proposed algorithm

Substring-Group Feature Extracting

The proposed algorithm utilizes a unique substring-group feature extraction process, which involves the following steps:

- Step (a) involves constructing a suffix tree using all strings from both the training and unlabeled test documents. This is done through the use of Ukkonen's algorithm, which has a time complexity of $O(n)$, where n is the number of characters in the text corpus. This step incorporates transductive learning.
- To extract key-nodes from the constructed suffix tree in step (b), the following is done. The suffix tree has m leaf-nodes and at most $m - 1$ internal nodes for an m character text corpus [34]. Since the text corpus's length m is typically very large, it is necessary to extract key-nodes from the $(2m - 1)$ nodes. In this study, the key-node extracting criteria proposed in [32] is employed, with recommended values of $L = 20$, $H = 8000$, $B = 8$, $P = 0.8$ and $Q = 0.8$. This ensures that the key-nodes are efficiently extracted from the suffix tree, thereby improving the overall efficiency and effectiveness of the algorithm.
- In step (c), each document's suffixes are matched with the suffix tree, and the matched key-nodes' IDs are taken as the content of the corresponding document.
- In step (d), unique key-node IDs are extracted as features for sentiment classification from the training section of the converted documents. This step ensures that the experiments' evaluation is an open test.
- Finally, in step (e), all converted documents are translated into corresponding vectors using the unique feature table produced by step (d).

As per the definition of suffix tree [34], each node in the tree corresponds to a substring-group of the text corpus. Hence, the IDs of the extracted key-nodes are referred to as substring-group features in this study. Additionally, it is worth noting that the time complexity of the computing steps a, b, and c is linear, as proven in [32]. This underscores the efficiency and scalability of the proposed algorithm for extracting and selecting substring-group features for sentiment classification.

Term Weighting

Standard text classification has traditionally relied on term frequency, but Pang et al. References [29] demonstrated that presence can yield better performance. As a result, this section employs both term presence ("bool," "three") and term frequency ("tf" and "$tfidf - c$"). The "$tfidf - c$" variant of the standard "$tfidf$" approach is used in this study, as it is widely used in text classification [35, 38]. The four term weighting approaches adopted in this study are defined as Eqs. (3.38), (3.9), and (3.40). By leveraging these term weighting methods, the proposed algorithm is able to effectively capture and represent the key features of the input text data, improving the accuracy and robustness of the sentiment classification process.

$$\text{bool}: \begin{cases} 1 & \text{if } tf\left(t_k, d_j\right) > 0 \\ 0 & \text{if } tf\left(t_k, d_j\right) = 0 \end{cases} \tag{3.38}$$

$$\text{three}: \begin{cases} 2 & \text{if } tf\left(t_k, d_j\right) > 1 \\ 1 & \text{if } tf\left(t_k, d_j\right) = 1 \\ 0 & \text{if } tf\left(t_k, d_j\right) = 0 \end{cases} \tag{3.39}$$

$$\text{tfidf}-c: \quad tf: \left(t_k, d_j\right) \quad \frac{tf\left(t_k, d_j\right) \times \log\frac{N}{df\left(t_k\right)}}{\displaystyle\sum_{t \in d_j}\left(tf\left(t_k, d_j\right) \times \log\frac{N}{df\left(t_k\right)^2}\right)} \tag{3.40}$$

In the proposed algorithm, each distinct term t_k corresponds to a single feature. The term frequency of t_k in document d_j is represented as $tf(t_k, d_j)$, while $df(t_k)$ denotes the number of documents in which the term t_k occurs. N represents the total number of training documents. By utilizing these measures, the algorithm is able to weigh the importance of each term in the input text corpus, allowing for more accurate and effective sentiment classification.

Feature Selecting

Document frequency (DF) is a measure of how frequently a term occurs across a set of documents. It is a straightforward and effective criterion for feature selection, with linear computation complexity that makes it readily scalable to large datasets. DF has been widely used in text categorization as a simple but effective feature selection method [39]. In this study, DF is leveraged to identify the discriminating substring-group features for both training and classification, enhancing the accuracy and efficiency of the sentiment classification process.

To calculate the document frequency (DF) of each feature in the training corpus, we count the number of documents in which each feature appears. The top N features with the highest DF scores are then selected for use in sentiment classification. This approach is based on the assumption that rare features are either noninformative for class prediction or have little influence on overall performance. By selecting only the most informative and influential features, the algorithm is able to achieve higher accuracy and efficiency in sentiment classification.

Classifying

In this stage, a machine learning algorithm is employed to train the sentiment classifier and predict the classifications of unlabeled test documents. Given the outstanding performance of support vector machines (SVMs) in sentiment classification [26, 27, 29, 30, 36, 40], SVMs are utilized in this study. The SVM^{light} package is used for both training and testing, with default parameters. By leveraging the power of SVMs, the proposed algorithm is able to accurately and efficiently classify sentiment in text data, enabling more effective analysis and understanding of large-scale text corpora.

3.7.3 Experimental Setup

Datasets

To evaluate the effectiveness and robustness of the proposed algorithm, it was tested on three publicly available datasets in three distinct languages: Chinese, English, and Spanish. Table 3.19 provides a brief overview of these datasets, which were used to assess the algorithm's performance across a range of linguistic and cultural contexts. By conducting experiments on these diverse datasets, we were able to demonstrate the wide applicability and effectiveness of the proposed algorithm in sentiment classification tasks.

The dataset of 160,000 Chinese hotel reviews is sourced from one of China's most popular websites for hotel and flight reservations. The "English_1400" dataset is widely recognized and frequently used for sentiment classification in English. The Spanish corpus comprises 400 reviews across a variety of categories, including cars, hotels, washing machines, books, cell phones, music, computers, and movies. For each category, there are 50 positive and 50 negative reviews, with the sentiment polarity being defined based on the number of stars given by the reviewers. By utilizing these diverse datasets, the proposed algorithm is able to achieve high accuracy and robustness in sentiment classification across multiple languages and domains.

To facilitate comparison with related works on the aforementioned datasets, different cross-validation methods were employed in the experiments. By employing different cross-validation methods, we were able to obtain a more comprehensive and reliable assessment of the algorithm's effectiveness and robustness in sentiment classification across multiple languages and domains.

Table 3.19 The summary of the open datasets

Language	Positive	Negative	n-fold CV	Encoding
Chinese_16,000	8000	8000	4	GB2312
English_1400	700	700	3	ASCII
Spanish_400	200	200	3	ISO-8859-2

Evaluation Metrics

In order to assess the effectiveness of the proposed algorithm for sentiment classification, we utilized the traditional evaluation metric of accuracy, which is commonly used in text categorization [38]. Furthermore, we also computed microF1 and macroPrecision metrics to enable a more comprehensive comparison with related works. By employing these multiple evaluation metrics, we were able to obtain a more nuanced and detailed understanding of the algorithm's performance and strengths in sentiment classification tasks.

3.7.4   Experiments

Comparisons

In order to compare the proposed algorithm with existing methods, we have listed the best performance achieved by typical methods on each dataset in Table 3.20. The "Number of Features" column represents the number of features used to achieve the best performance. By comparing the proposed algorithm's performance with existing methods, we were able to demonstrate the effectiveness and superiority of the proposed approach in sentiment classification across multiple languages and domains.

Table 3.20 shows that the proposed algorithm (highlighted in gray) achieves superior performance on three different language datasets, despite not utilizing any preprocessing steps such as word segmentation or stemming. Notably, the proposed algorithm processes all datasets in a uniform manner, without employing different language-specific approaches. Another observation is that the proposed algorithm uses a larger number of features compared to most other algorithms, indicating that its promising performance comes at the cost of high feature dimensionality. Nonetheless, the superior accuracy and robustness of the proposed algorithm in sentiment classification tasks demonstrate its potential for a wide range of practical applications.

Table 3.20 Comparisons with the best performance of the existing typical methods

Language	Techniques	Best performance (%)	#Features
Chinese (16,000)	SVM(word bigrams, tfidf-c) [27]	$91.2^{microF1}$	251,289
	SVM(word bigrams, tfidf-c) [27]	$91.6^{microF1}$	128,049
	SVM(key substring-groups + DF, tfidf-c)	$\mathbf{94.0^{microF1}}$	41,454
English (1400)	SVM(character unigrams, bool) [30]	$82.9^{accuracy}$	16,165
	SVM(key substring-groups + DF, tfidf-c)	$\mathbf{84.3^{accuracy}}$	28,726
Spanish (400)	No existing work has used this corpus yet		
	SVM(key substring-groups + DF, tfidf-c)	$\mathbf{78.7^{accuracy}}$	2519

Multilingual Characteristics

One of the key strengths of the proposed algorithm is its ability to treat input documents as character sequences, without requiring any word segmentation technology. This means that the algorithm is not dependent on the syntax or semantic structures of the input text, making it suitable for use with any language in any encoding format. This has been demonstrated by the experiments presented in Table 3.20, which show the algorithm's superior performance on datasets in different languages. By leveraging its language-independent approach, the proposed algorithm offers a flexible and adaptable solution for sentiment classification that can be applied to a wide range of text data sources.

In addition to its language-independent approach, the proposed algorithm is also capable of handling text corpora containing both English and Chinese words simultaneously. To demonstrate this capability, we conducted an experiment on a mixed-language dataset comprising the "English_1400" corpus and 1400 randomly selected Chinese reviews (700 pos + 700 neg) from the "Chinese_16,000" corpus. Three-fold cross-validation was employed, and the results are presented in Fig. 3.19. As shown in the Figure, the proposed algorithm achieved promising performance (represented by the dark blue curve) on the mixed-language dataset, surpassing the performance achieved when using only the English corpus. This demonstrates the algorithm's ability to effectively handle mixed-language text data sources, enhancing its practical applicability in real-world settings.

Feature Frequency Versus Feature Presence

The performance of sentiment classification is heavily influenced by the text representation used. To demonstrate the impact of different term weighting approaches on the performance of the sentiment classifier, we conducted a series of experiments. By

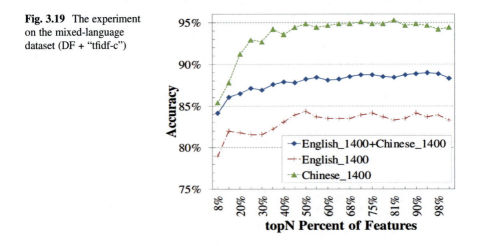

Fig. 3.19 The experiment on the mixed-language dataset (DF + "tfidf-c")

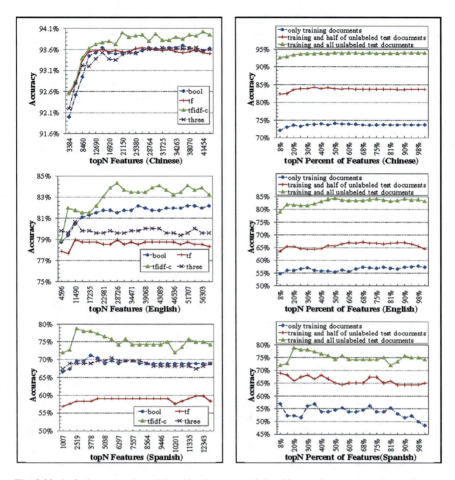

Fig. 3.20 Left: the accuracies achieved by the proposed algorithm on three open sentiment datasets in different languages. Right: comparisons of transductive learning (green and red) and inductive learning (dark blue)

comparing the results obtained using different weighting methods, we were able to identify which approach is most effective for accurately capturing and representing the key features of the input text data. These experiments highlight the importance of carefully selecting and optimizing the term weighting approach in sentiment classification tasks, as it directly impacts the accuracy and robustness of the resulting model.

Figure 3.20 (left) illustrates the impact of different term weighting approaches on the performance of the sentiment classifier. The results show that the choice of term weighting method has a significant effect on the performance of the classification model. Among all the term weighting approaches tested, "tfidf-c" consistently outperformed all other methods on the three open datasets in different languages, while "tf" performed the worst. Additionally, "bool" consistently achieved better performance than "three." These results highlight the importance of carefully selecting the appropriate term weighting approach in sentiment classification, as it significantly impacts the accuracy and effectiveness of the resulting model.

These findings are consistent with Pang's research, which suggested that superior performance is achieved by considering only feature presence ("bool"), rather than feature frequency ("tf") [30]. However, the advanced feature weighting method of "tfidf-c" was found to be superior to "bool" in the proposed algorithm. As a result, "tfidf-c" was used in all subsequent experiments in the following subsections. These results reinforce the importance of selecting an appropriate term weighting approach for sentiment classification, as it significantly impacts the accuracy and effectiveness of the resulting model.

Influence of Feature Selecting

The left side of Fig. 3.20 also demonstrates the effectiveness of feature selection in sentiment classification. The DF-based feature selection method was found to be capable of eliminating up to 50% or more of the unique substring-group features, without sacrificing or even improving classification accuracy. This is particularly evident in the "tfidf-c" curves (shown in green). Furthermore, Table 3.20 shows that all the best performances achieved by the proposed algorithm utilized DF-based feature selection methods. These results highlight the importance and effectiveness of feature selection in sentiment classification, as it enables the algorithm to focus on the most informative and influential features in the input text data, enhancing the accuracy and efficiency of the classification process.

Drawing upon the observations made above, we can conclude that the extracted substring-group features in step 1 may contain redundancy, and that feature selection methods should be employed to eliminate this redundancy and optimize the feature set. By eliminating redundant features, the algorithm can focus on the most informative and influential features in the input text data, thereby enhancing the accuracy and efficiency of the sentiment classification process. Therefore, feature selection is a critical step in the proposed algorithm, and should be carefully considered and optimized to achieve the best possible performance.

Transductive Learning Vs. Inductive Learning

The experiments conducted below demonstrate the effectiveness of employing a transductive learning-based algorithm, as opposed to traditional inductive methods. The right side of Fig. 3.20 presents the experimental results obtained on the three open datasets in three different languages. By comparing the performance of the transductive learning-based approach with traditional inductive methods, we were able to demonstrate the superior accuracy and robustness of the former in sentiment classification tasks. These results highlight the potential of transductive learning as a powerful and effective approach for sentiment classification in large-scale text data sources.

The results presented in the right side of Fig. 3.20 demonstrate that the transductive learning-based algorithm (represented by the green and red curves) is highly effective for sentiment classification tasks. As the number of unlabeled test documents included in the suffix tree construction grows, the performance of the transductive learning-based approach improves significantly. These findings highlight the potential of transductive learning as a powerful and effective approach for sentiment classification, particularly in scenarios where large amounts of unlabeled data are available. By leveraging the vast amounts of unlabeled data available in many real-world applications, the proposed approach offers a scalable and efficient solution for sentiment classification in large-scale text data sources.

Additional interesting observations can be made by examining the data presented in the white background of Table 3.20 and the dark blue curves in the right side of Fig. 3.20. Notably, the substring-group-based algorithms in the inductive learning setting were found to be inferior to algorithms utilizing character or word Ngrams features. This highlights the importance of transductive learning in sentiment classification from another perspective, as it enables the algorithm to leverage the vast amounts of unlabeled data available to improve the accuracy and robustness of the classification model. These findings further reinforce the potential of transductive learning as a powerful and effective approach for sentiment classification in large-scale text data sources.

The improvement in performance achieved by transductive learning can be attributed to the fact that as more unlabeled test documents are included in the suffix tree construction, the structure of the tree becomes more complete and representative of the text corpus. This enables the extraction of more informative and representative substring-group features from the suffix tree, resulting in more accurate and representative converted documents. In this way, the structural information contained within the unlabeled test documents indirectly contributes to the feature subsumption for sentiment classification, enhancing the accuracy and robustness of the resulting model. These findings highlight the potential of transductive learning as a powerful and effective approach for sentiment classification in large-scale text data sources, particularly in scenarios where vast amounts of unlabeled data are available.

3.7.5 Conclusion

This study incorporates both feature extraction and feature selection methods into sentiment classification, and investigates the synergistic effect of these techniques. Additionally, the proposed algorithm combines substring-group features with transductive learning, resulting in a powerful and effective approach for sentiment classification in large-scale text data sources. By leveraging the advantages of both feature extraction and selection techniques, the proposed algorithm is able to identify and utilize the most informative and influential features in the input text data, enhancing the accuracy and efficiency of the sentiment classification process. Moreover, by combining substring-group features with transductive learning, the algorithm is able to effectively leverage vast amounts of unlabeled data, further enhancing its accuracy and robustness. These findings demonstrate the potential of the proposed algorithm as a powerful and effective solution for sentiment classification in a wide range of practical applications.

The proposed algorithm was evaluated through experiments conducted on open datasets in three different languages: Chinese, English, and Spanish. These experiments demonstrated that the proposed algorithm achieved superior performance compared to existing algorithms, without the need for any preprocessing steps such as word segmentation or stemming. Furthermore, the proposed algorithm was found to be multilingual, and can be used directly for sentiment classification with any language in any encoding format. The "tfidf-c" term weighting approach was found to perform best in the proposed algorithm. Additionally, experimental results highlighted the significant improvement in classifier performance achieved through the incorporation of structural information from unlabeled test documents using the transductive learning-based algorithm. These findings demonstrate the potential of the proposed algorithm as a powerful and effective solution for sentiment classification in a wide range of practical applications, offering a scalable and adaptable solution for sentiment analysis in large-scale text data sources.

Moving forward, we plan to analyze the misclassifications made by the classifier to gain insights into how it can be further improved. We also intend to explore additional feature extraction methods to enhance the overall performance of sentiment classification. By identifying and utilizing the most informative and influential features in the input text data, we aim to improve the accuracy and efficiency of the sentiment classification process, and further enhance the practical applicability of the proposed algorithm in real-world settings. These efforts will help to advance the state-of-the-art in sentiment classification, and enable more effective and efficient analysis of large-scale text data sources.

3.8   Summary

This chapter offers a comprehensive exploration of opinion mining and sentiment analysis, presenting innovative methods to address diverse challenges in these domains. This chapter introduces unsupervised approaches for identifying evaluative sentences, proposes techniques for effective feature grouping, and enhances topic modeling with constraints. It evaluates the proposed methods using real-life datasets, demonstrating their superiority over baseline approaches. Additionally, this chapter evaluates various feature types for sentiment classification and introduces a lexicon-based framework for sentiment analysis. It further incorporates feature extraction, selection, and transductive learning, showcasing their synergistic effects in sentiment classification. Overall, this chapter provides valuable insights, practical solutions, and directions for future research in opinion mining and sentiment analysis.

References

1. Yu H, Hatzivassiloglou V (2003) Towards Answering Opinion Questions: Separating Facts from Opinions and Identifying the Polarity of Opinion Sentences. Proceedings of the 2003 Conference on Empirical Methods in Natural Language Processing, 129–136
2. Wilson T, Wiebe J, Hwa R (2004) Just how mad are you? finding strong and weak opinion clauses. Proceedings of the 19th National Conference on Artifical Intelligence, 761–767
3. Wiebe J, Riloff E (2005) Creating subjective and objective sentence classifiers from unannotated texts. Proceedings of the 2005 International Conference on Intelligent Text Processing and Computational Linguistics, 3406: 486–497
4. Kim SM, Hovy EH (2006) Identifying and analyzing judgment opinions. Proceedings of the main conference on Human Language Technology Conference of the North American Chapter of the Association of Computational Linguistics, 200–207
5. Qiu G, Liu B, Jiajun B et al (2011) Opinion word expansion and target extraction through double propagation. Comput Linguist 37(1):9–27
6. Kleinberg JM (1999) Authoritative sources in a hyperlinked environment. J ACM 46(5): 604–632
7. Andrzejewski D, Zhu X, Craven M (2009) Incorporating domain knowledge into topic modeling via Dirichlet forest priors. Proceedings of the 26th Annual International Conference on Machine Learning, 25–32
8. Wagstaff K, Cardie C, Rogers S, et al (2001) Constrained k-means clustering with background knowledge. Proceedings of the 18th International Conference on Machine Learning, 577–584
9. Andrzejewski D, Zhu X (2009) Latent Dirichlet allocation with topic-in-set knowledge. Proceedings of the North American Chapter of the Association for Computational Linguistics 2009 Workshop on Semi-Supervised Learning for Natural Language Processing, 43–48
10. Nigam K, Mccallum AK, Thrun S (2000) Text classification from labeled and unlabeled documents using EM[J]. Mach Learn 39:103–134
11. Jonathan C, David B (2009) Relational topic models for document networks[J]. J Mach Learn Res 5:81–88
12. Griffiths TL, Steyvers M (2002) A probabilistic approach to semantic representation. Proceedings of the 24th Annual Conference of the Cognitive Science Society, 381–386
13. Griffiths TL, Steyvers M (2003) Prediction and semantic association. Proceedings of the 15th International Conference on Neural Information Processing Systems, 11–18
14. Clemons EK, Gao G, Hitt LM (2006) When online reviews meet hyperdifferentiation: A study of the craft beer industry. Proceedings of the 39th Annual Hawaii International Conference on System Sciences, 23(2): 149–171
15. Carenini G, Ng RT, Zwart E (2005) Extracting knowledge from evaluative text. Proceedings of the 3rd international conference on Knowledge capture, 11–18
16. Rand WM (1971) Objective criteria for the evaluation of clustering methods. J Am Stat Assoc 66(336):846–850
17. Branavan SRK, Harr C, Eisenstein J, et al (2008) Learning document-level semantic properties from free-text annotations. Proceedings of Association for Computational Linguistics, 569–603
18. Cardie C, Wagstaff K (1999) Noun phrase coreference as clustering. Proceedings of the 1999 Joint SIGDAT Conference on Empirical Methods in Natural Language Processing and Very Large Corpora, 82–89
19. Griffiths TL, Steyvers M (2004) Finding scientific topics. Proc Natl Acad Sci 101:5228–5535

20. Guo H, Zhu H, Guo Z, et al. (2009) Product feature categorization with multilevel latent semantic association. Proceedings of the 18th Association for Computing Machinery conference on Information and knowledge management, 1087–1096
21. Jiang JJ, Conrath DW (1997) Semantic similarity based on corpus statistics and lexical taxonomy. Proceedings of the 10th Research on Computational Linguistics International Conference, 19–33
22. Philip R (1995) Using Information Content to Evaluate Semantic Similarity in a Taxonomy. Proceedings of the 14th International Joint Conference on Artificial Intelligence, 1:448–453
23. Lin D (1998) An information-theoretic definition of similarity. Proceedings of the 15th International Conference on Machine Learning, 296–304
24. Pang B, Lee L (2008) Opinion mining and sentiment analysis. Found Trends Inf Retr 2(1–2): 1–135
25. Riloff E, Patwardhan S, Wiebe J (2006) Feature subsumption for opinion analysis. Proceedings of the 2006 Conference on Empirical Methods in Natural Language Processing, 440–448
26. Tan S, Zhang J (2008) An empirical study of sentiment analysis for Chinese documents. Exp Syst Appl 34(4):2622–2629
27. Li J, Sun M (2007) Experimental study on sentiment classification of chinese review using machine learning techniques. Proceedings of Natural Language Processing and Knowledge Engineering, 393–400
28. Raaijmakers S, Kraaij W (2008) A shallow approach to subjectivity classification. Proceedings of the 2nd International AAAI Conference on Weblogs and Social Media, 2(1): 216–217
29. Pang B, Lee L (2004) A sentimental education: Sentiment analysis using subjectivity summarization based on minimum cuts. Proceedings of the 42nd Annual Meeting of the Association for Computational Linguistics, 271–278
30. Pang B, Lee L, Vaithyanathan S (2002) Thumbs up? Sentiment classification using machine learning techniques. Proceedings of the 2002 Association for Computational Linguistics conference on Empirical methods in natural language processing, 79–86
31. Indurkhya N, Damerau FJ (2010) Handbook of natural language processing, 2nd edn. Chapman & Hall/CRC, London
32. Zhang D, Lee WS (2006) Extracting key-substring-group features for text classification. Proceedings of the 12th Association for Computing Machinery Special Interest Group on Knowledge Discovery and Data Mining international conference on Knowledge discovery and data mining, 474–483
33. Ukkonen E (1995) On-line construction of suffix trees. Algorithmica 14(3):249–260
34. Gusfield D (1997) Algorithms on strings, trees, and sequences. Cambridge University Press, Cambridge
35. Joachims T (1997) A probabilistic analysis of the Rocchio algorithm with TFIDF for text categorization. Proceedings of the 14th International Conference on Machine Learning, 143–151
36. Joachims T (1998) Text categorization with support vector machines: Learning with many relevant features. Proceedings of the 1998 European Conference on Machine Learning, 1398: 137–142
37. Liu B (2011) Web data mining; exploring hyperlinks, contents, and usage data. Springer Publishing Company, Berlin
38. Sebastiani F (2002) Machine learning in automated text categorization. ACM Comput Surv 34(1):1–47
39. Yang Y, Pedersen JO (1997) A comparative study on feature selection in text categorization. Proceedings of the 14th International Conference on Machine Learning, 412–420
40. Mullen T, Collier N (2004) Sentiment analysis using support vector machines with diverse information sources. Proceedings of the 2004 Conference on Empirical Methods in Natural Language Processing, 412–418

Chapter 4
Unimodal Sentiment Analysis

Abstract This chapter explores the realm of sentiment analysis, covering diverse domains such as text, audio, and facial expressions. It introduces novel approaches that address the limitations of existing methods, emphasizing the significance of semantic relationships, effective fusion of heterogeneous features, and the benefits of multitask learning. The proposed techniques, including word2vec and SVM^{perf} for sentiment classification, RCMSA and CHFFM for audio sentiment analysis, and CMCNN with CCAM and SCAM for facial expression recognition, exhibit superior performance in their respective domains. This chapter sets the stage for the book, showcasing innovative methods that advance the field of sentiment analysis and provide valuable insights for researchers and practitioners alike.

4.1 Text Sentiment Analysis Based on Word2vec and SVM^{perf}

4.1.1 Introduction

Machine learning-based approaches for sentiment classification have become increasingly popular due to their outstanding performance. However, most existing research in this area focuses on the extraction of lexical and syntactic features, while the semantic relationships between words are often ignored. To address this limitation, we propose a novel method for sentiment classification based on word2vec and SVM^{perf}, which is capable of capturing the semantic features of text data. Our research consists of two main parts. Firstly, we use word2vec to cluster similar features, demonstrating its ability to capture semantic relationships within a selected domain and the Chinese language. Next, we train and classify comment texts using word2vec and SVM^{perf}, incorporating lexicon-based and part-of-speech-based feature selection methods to generate the training file. Our experiments were conducted on a dataset of Chinese comments on clothing products, and the results demonstrate the superior performance of our method in sentiment classification. These findings highlight the potential of our approach as a powerful and effective solution for sentiment classification, capable of capturing and utilizing the semantic relationships between words in large-scale text data sources.

4.1.2   Methodology

Previous research has demonstrated that word2vec is highly effective in text classification and clustering in English [1–3]. However, it remains unclear whether word2vec can provide similar performance benefits for Chinese text clustering. To investigate this question, we first use word2vec to group synonyms together under the same feature groups. We then leverage the combination of word2vec and SVM^{perf} to classify comment texts into positive and negative classes. The general framework of our approach is illustrated in Fig. 4.1. In the following subsections, we provide a detailed overview of these two steps, highlighting the key techniques and methodologies employed in our approach. By leveraging the power of word2vec and SVM^{perf}, we aim to develop a highly effective and efficient solution for sentiment classification in Chinese text data.

Similar Features Clustering

To produce a useful comment summary, it is important to group different words expressing the same product feature under the same feature group [4, 5]. In our approach, we use word2vec to cluster similar features. The clustering process involves three main steps:

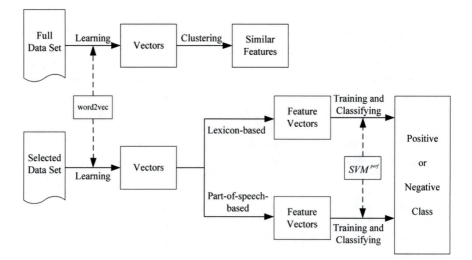

Fig. 4.1 The general framework of our work

Table 4.1 The key parameters of training command

Parameters	Explanations	Default values
Train	Name of input file	Train.txt
Output	Name of output file	Vectors.bin
cbow	Choice of training model 0: Skip-gram model 1: CBOW model	0
Size	Dimension of vectors	200
Window	Size of training window	5
Negative	Choice of training method 0: Hierarchical Softmax method 1: Negative sampling method	0
hs	Choice of training method 0: Negative sampling method 1: Hierarchical Softmax method	1
Sample	Threshold of sampling	1e 3
Threads	Number of running threads	12
Binary	Mode of storage 0: Common format 1: Binary format	1

1. **Step 1. Preprocessing**: In our approach for sentiment classification in Chinese text data, we utilize the ICTCLAS system, which has been developed by the Institute of Computing Technology at the Chinese Academy of Science. This system enables us to segment the collected Chinese comment texts into individual words and assign them appropriate part-of-speech (POS) tags. After removing stop words, punctuation characters, and other unnecessary elements, we obtain a clean and processed training file. This preprocessing step is essential for accurate and efficient sentiment classification, as it ensures that the input text data is properly formatted and annotated, and free from irrelevant or redundant information. By leveraging the capabilities of the ICTCLAS system, we can enhance the quality and reliability of our sentiment analysis results, and obtain more insightful and informative insights from large-scale Chinese text data sources.

2. **Step 2. Training**: Our approach involves training a model file using word2vec. We use the training file as input, and the word2vec tool constructs a vocabulary from the file and learns high-dimensional vector representations of words based on the specified parameters. Table 4.1 provides detailed explanations and default values of the key parameters used in the training command. The output of this process is a model file that captures the semantic relationships between words in the training data. By leveraging the power of word2vec to generate high-quality word embeddings, we can enhance the accuracy and efficiency of sentiment classification in Chinese text data. This approach enables us to effectively capture the nuances and complexities of language use in large-scale text data sources, and to produce meaningful and informative insights that can inform decision-making and strategy development in various domains.

3. **Step 3. Clustering**: Our approach for clustering similar features using word2vec involves leveraging the model file generated during training. If the model file is stored in the common format, we can access each word in the document and its corresponding vector. We use the "distance" command provided by word2vec to compare the semantic similarity between words and cluster synonyms. This command reads every word and its corresponding vectors in the model file, and calculates the semantic distances between the input word and other words using cosine similarity. The resulting cosine values indicate the degree of semantic similarity between each pair of words. We sort these values in descending order to obtain a list of the closest words to the input word and their corresponding distances. This approach allows us to effectively group together synonyms and related features, enhancing the accuracy and efficiency of sentiment classification in Chinese text data. By leveraging the power of word2vec to identify and cluster similar features, we can extract deeper insights from large-scale text data sources, and develop more effective strategies for various applications.

Sentiment Classification

In our approach to sentiment classification, we employ a combination of two tools, word2vec and SVM^{perf}, which differs from traditional methods. To begin, we use word2vec to prune words whose occurrence frequency in the corpus is less than five before training. The remaining words are considered frequent and treated as candidate features. Word2vec then trains the corpus and generates a model file that contains a list of frequent words and their corresponding vector representations, each with a single dimension. We use two feature selection methods, one based on lexicon and the other based on part-of-speech, to identify valuable features from the candidate set. By leveraging the power of word2vec and SVM^{perf}, our approach allows us to effectively classify sentiment in Chinese text data, enhancing the accuracy and efficiency of sentiment analysis. This approach enables us to identify relevant features and capture the semantic relationships between words, providing deeper insights into the underlying sentiment of large-scale text data sources.

1. **Feature selection method based on lexicon**: Our approach to sentiment classification relies on a comprehensive dictionary of sentiment words and phrases, including their associated orientations and weights, as well as intensifiers and negations [6]. For this purpose, we have selected the HowNet sentiment word set and IARDict as our two original dictionaries. The HowNet sentiment word set consists of a list of opinion words extracted from the HowNet online knowledge base, while IARDict was collected by our research team in prior work. By leveraging these dictionaries, we can accurately identify sentiment words and phrases in Chinese text data, and assign them appropriate orientations and weights based on their semantic context. This approach enables us to perform sentiment classification at scale, providing valuable insights into the underlying sentiment of large-scale text data sources in various domains.

To enhance the coverage and accuracy of our sentiment classification approach, we have implemented a method for expanding our original dictionaries. Specifically, we begin by extracting the top ten opinion words with the highest weight from the original dictionaries, which serve as input words. We then use the "distance" command to identify additional opinion words that are semantically similar to the input words. By using this approach, we can expand our original dictionaries and capture a wider range of sentiment words and phrases in Chinese text data. This method enables us to perform sentiment classification at scale, providing valuable insights into the underlying sentiment of large-scale text data sources in various domains.

Algorithm 4.1: Feature selection based on lexicon

1: word_set ← frequent words
2: dic_set ← opinion words in lexicon
3: **for** each *w* in dic_set **do**
4:      **if** *w* is in word_set **then**
5:          add *w* to feature_set
6:      **else**
7:          continue
8:      **end if**
9: **end for**

After expanding our dictionaries, we select the final training features by extracting words that appear in both the candidate feature set and the expanded dictionaries. This approach ensures the selection of relevant and accurate features for sentiment classification in Chinese text data. Algorithm 4.1 summarizes our feature selection process, resulting in a final feature set used to train our sentiment classification model. Our comprehensive approach, leveraging various features and dictionaries, enables us to perform sentiment classification at scale and provide valuable insights into the underlying sentiment of large-scale text data sources in various domains.

2. **Feature selection method based on part-of-speech**: Our feature selection method for sentiment classification involves selecting valuable features based on the part-of-speech of words. However, it is important to note that different choices of part-of-speech can lead to varying results [7]. For example, if only adjectives are chosen as features, the classification result may not be better than if adverbs, verbs, and adjectives are selected together. This is due to the fact that many words with different parts-of-speech can serve as sentiment indicators. By considering a range of part-of-speech categories for feature selection, our approach ensures that we capture the full range of sentiment indicators in Chinese text data. This enhances the accuracy and efficiency of sentiment classification and provides valuable insights into the underlying sentiment of large-scale text data sources in various domains.

Our feature selection method for sentiment classification involves selecting the most common words in the documents based on their part-of-speech. Specifically, we keep adjectives, adverbs, verbs, and nouns, which are among the most common parts-of-speech in Chinese text data. We then generate different combinations of these parts-of-speech as training features. By leveraging a comprehensive set of features based on the most common parts-of-speech categories, our approach enables us to accurately capture the full range of sentiment indicators in Chinese text data. This enhances the accuracy and efficiency of sentiment classification and provides valuable insights into the underlying sentiment of large-scale text data sources in various domains.

3. **Training and classifying**: In this stage, we use a classifier to train the selected feature vectors and predict the sentiment polarity of the test documents. By leveraging the power of machine learning techniques, we can accurately classify sentiment in Chinese text data, enhancing the accuracy and efficiency of sentiment analysis. The classifier is trained on the selected feature vectors to identify patterns and relationships between the features and the sentiment polarity of the documents. This enables us to generate accurate predictions for the sentiment polarity of new, unseen documents. Our approach enables us to perform sentiment classification at scale, providing valuable insights into the underlying sentiment of large-scale text data sources in various domains.

   Numerous studies have demonstrated the superior performance and robustness of SVM compared to other state-of-the-art models for classification tasks [8, 9]. As a result, we have adopted SVM in this section for our sentiment classification approach. To further optimize the performance of SVM, we have employed SVM^{perf}, an optimized version of SVM^{light} developed by Joachims, who is also the author of SVM^{light}. Although the overall architecture of SVM^{perf} follows that of SVMlight, the kernel algorithm is more advanced, resulting in enhanced speed and accuracy of classification. We utilize the SVM^{perf} package for training and testing, enabling us to accurately classify sentiment in Chinese text data at scale. By leveraging the power of SVM^{perf}, we can effectively capture the nuances and complexities of language use in large-scale text data sources, and produce meaningful and informative insights that can inform decision-making and strategy development in various domains.

4.1.3   Experiments

Data Sets

Our approach begins by crawling over 100,000 Chinese comments on clothing products from Amazon. These comments are then processed to remove duplicates and meaningless data, resulting in a final dataset of 92,220 comments.

In our study, we conducted two experiments on datasets of varying sizes. Our approach to similar features clustering based on word2vec does not require identifying the polarity of texts, making it a versatile and efficient method for sentiment analysis. Additionally, the performance of word2vec improves with larger datasets. Therefore, we utilized all collected comments as the dataset for similar features

Table 4.2   A brief summary of data sets

Our work	Positive	Negative	Total number
Similar features clustering	N/A	N/A	92,220
Sentiment classification	5000	5000	10,000

clustering, enabling us to accurately capture the nuances and complexities of language use in Chinese text data at scale. By leveraging a comprehensive dataset and advanced clustering techniques, our approach provides valuable insights into the underlying sentiment of large-scale text data sources in various domains.

Our primary focus is on sentiment classification using word2vec and SVM^{perf}, a supervised machine learning method. To accomplish this, we divided all collected comments into five levels based on user star ratings. Initially, we treated comments with a 5-star rating as positive class and comments with a 1-star rating as negative class. However, the ratio of positive class to negative class was highly imbalanced, with a nine-to-one ratio. This imbalance negatively impacted the classification results. To address this issue, we adjusted our strategy. All comments with a 1-star rating were still considered the negative class, while an equal number of randomly selected comments with a 5-star rating were treated as the positive class. Table 4.2 provides a brief summary of the resulting dataset. To conduct our experiments, we divided the dataset into two equal parts, with 2500 positive comments and 2500 negative comments used for training, and the remaining half used for testing. By utilizing a balanced dataset and advanced machine learning techniques, our approach enables us to accurately classify sentiment in Chinese text data at scale, providing valuable insights into the underlying sentiment of large-scale text data sources in various domains.

Evaluation Criteria

To evaluate the effectiveness of our sentiment classification approach, we utilized precision, recall, and F1 score, which are classic metrics commonly used in classification tasks. These metrics were used to measure the performance of both the positive and negative classes. In addition to these metrics, we also used accuracy as a criterion to evaluate the overall performance of our sentiment classification approach. By leveraging a comprehensive set of evaluation metrics, our approach enables us to accurately classify sentiment in Chinese text data at scale, providing valuable insights into the underlying sentiment of large-scale text data sources in various domains.

Experimental Results

To assess the effectiveness of our similar features clustering and sentiment classification approaches, we conducted experiments and evaluated the results. This section presents and discusses the outcomes of our experiments. By utilizing advanced techniques and metrics for evaluation, our approach enables us to accurately classify sentiment in Chinese text data at scale, providing valuable insights into the underlying sentiment of large-scale text data sources in various domains.

1. **Result of similar features clustering**: This subsection focuses on presenting and analyzing the results of our similar features clustering approach. To perform similar features clustering on the dataset of Chinese clothing comments, we selected four representative features: Jia4Ge2 (price), Mian4Liao4 (fabric), Chi3Ma3 (size), and Kuan3Shi4 (style). These features were chosen because they appeared most frequently in the Chinese clothing comments. We then obtained lists of synonyms and kept only the two-character nouns before selecting the top five words as the final clustering results. To optimize our approach, we trained word2vec at five different dimensions of vector. Table 4.3 displays the results of our similar features clustering approach based on word2vec. For each representative feature, we identified similar features that have the same or similar Chinese meaning. Interestingly, the lists of similar features remain consistent across different dimensions of vector, with only slight variations in the sequence of certain words. These minor variations have little impact on the accuracy of clustering. The nearly perfect results demonstrate the powerful capability of word2vec to capture the deep semantic relationships between words in Chinese text clustering.

2. **Performance of sentiment classification**: The proposed approach for sentiment classification is based on word2vec and SVM^{perf}, which adopts two feature selection methods, lexicon-based method and part-of-speech-based method. In this subsection, their classification performance is shown and discussed respectively.

Table 4.4 presents the performance of our lexicon-based feature selection method. We evaluated three different types of features and found that the combination of IARDict and HowNet performed slightly better than IARDict alone in terms of total accuracy. However, HowNet features had the worst performance overall. This could be due to the fact that HowNet features do not include negation words, which can alter the sentiment orientation and lead to incorrect classification results. The combination of IARDict and HowNet features, on the other hand, includes positive words, negative words, negation words, and intensifiers, taking into account not only the sentiment itself but also sentiment shifts. By leveraging a comprehensive set of lexicon-based features and machine learning techniques, our approach enables us to accurately classify sentiment in Chinese text data at scale, providing valuable insights into the underlying sentiment of large-scale text data sources in various domains.

Table 4.5 displays the performance of our part-of-speech-based feature selection method. Our analysis indicates that selecting adjectives, adverbs, and verbs as features outperforms other selection strategies in terms of F1 and accuracy. Conversely, selecting only adjectives and adverbs produces disappointing results, with the worst precision, recall, F1, and accuracy. Selecting all frequent words as features

Table 4.3 The results of similar features clustering based on word2vec

Representative features	Dimensions of vector	Similar features					
Jia4Ge2	200	Jia4Wei4	Jia4Qian2	Jia4Ma3	Jia4Zhi2	Tian1Jia4
	500	Jia4Wei4	Jia4Qian2	Jia4Ma3	Jia4Zhi2	Tian1Jia4
	1000	Jia4Wei4	Jia4Qian2	Jia4Ma3	Jia4Zhi2	Tian1Jia4
	5000	Jia4Wei4	Jia4Qian2	Jia4Ma3	Tian1Jia4	Jia4Zhi2
	10,000	Jia4Wei4	Jia4Qian2	Jia4Ma3	Jia4Zhi2	Tian1Jia4
Mian4Liao4	200	Liao4Zi	Bu4Liao4	Zhi4Di4	Cai2Liao4	Cai2Liao4
	500	Liao4Zi	Bu4Liao4	Cai2Liao4	Zhi4Di4	Cai2Zhi4
	1000	Liao4Zi	Bu4Liao4	Zhi4Di4	Cai2Zhi4	Yi1Liao4
	5000	Liao4Zi	Bu4Liao4	Zhi4Di4	Cai2Zhi4	Yi1Liao4
	10,000	Liao4Zi	Bu4Liao4	Zhi4Di4	Cai2Zhi4	Cai2Liao4
Chi3Ma3	200	Chi3Cun4	Hao4Ma3	Xing2Hao4	Ma3Hao4	Ma3Zi
	500	Chi3Cun4	Hao4Ma3	Ma3Zi	Ma3Hao4	Xing2Hao4
	1000	Chi3Cun4	Hao4Ma3	Xing2Hao4	Ma3Hao4	Ma3Zi
	5000	Chi3Cun4	Hao4Ma3	Xing2Hao4	Ma3Hao4	Ma3Zi
	10,000	Chi3Cun4	Hao4Ma3	Ma3Zi	Ma3Hao4	Xing2Hao4
Kuan3Shi4	200	Yang4Shi4	Kuan3Xing2	Wai4Guan1	Yang4Zi	Wai4Xing2
	500	Yang4Shi4	Kuan3Xing2	Yang4Zi	Wai4Xing2	Wai4Guan1
	1000	Yang4Shi4	Kuan3Xing2	Wai4Guan1	Yang4Zi	Wai4Xing2
	5000	Yang4Shi4	Kuan3Xing2	Yang4Zi	Wai4Guan1	Wai4Xing2
	10,000	Yang4Shi4	Kuan3Xing2	Yang4Zi	Wai4Guan1	Shi4Yang4

Table 4.4 The performance of sentiment classification based on lexicon

Features	Positive			Negative			Accuracy (%)
	Precision (%)	Recall (%)	F1 (%)	Precision (%)	Recall (%)	F1 (%)	
IARDict	91.36	86.70	88.97	87.35	91.80	89.52	89.25
HowNet	86.27	87.65	86.95	87.45	86.05	86.74	86.85
The combination	91.14	88.50	89.80	88.82	91.40	90.09	89.95

Table 4.5 The performance of sentiment classification based on part-of-speech

Features	Positive			Negative			Accuracy (%)
	Precision (%)	Recall (%)	F1 (%)	Precision (%)	Recall (%)	F1 (%)	
{a,d}	87.21	82.85	84.97	83.67	87.85	85.71	85.35
{a,d,v}	91.38	89.00	90.17	89.28	91.60	90.42	90.30
{a,d,n}	91.91	85.15	88.40	86.17	92.50	89.22	88.83
{a,d,v, n}	91.89	84.95	88.28	86.01	92.05	89.14	88.72
All	92.66	84.60	88.45	85.83	93.30	89.41	88.95

Table 4.6 Accuracy for different combinations with two feature selection methods

Method	Lexicon-based (%)	Part-of-speech-based (%)
TF-IDF + LibSVM	82.81	82.05
Word2vec + LibSVM	84.28	87.10
TF-IDF + SVM^{perf}	87.63	83.48
Word2vec + SVM^{perf}	89.95	90.30

results in high precision of the positive class and recall of the negative class, but lower recall of the positive class and precision of the negative class cause a decrease in F1 value and overall accuracy. The remaining two strategies perform similarly, with little difference in performance. By leveraging advanced part-of-speech tagging techniques and machine learning algorithms, our approach enables us to accurately classify sentiment in Chinese text data at scale, providing valuable insights into the underlying sentiment of large-scale text data sources in various domains.

To evaluate the effectiveness of our approach, we compared the performance of our TF-IDF weighting scheme and LibSVM classification model to other combinations. Table 4.6 presents the results of this comparison. As anticipated, our approach outperformed other combinations, demonstrating the effectiveness of our advanced techniques and algorithms.

In order to achieve the highest accuracy in sentiment classification, we fine-tuned our experiments by adjusting the regularization parameter C of SVM^{perf}. We selected the best strategy for each feature selection method, which included the

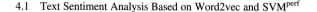

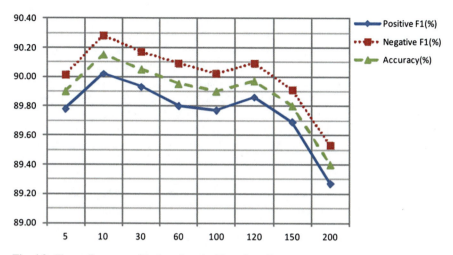

Fig. 4.2   The performance of lexicon-based with various C

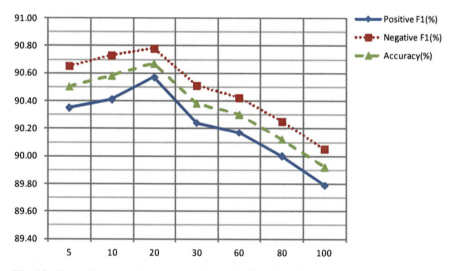

Fig. 4.3   The performance of part-of-speech-based with various C

combination of IARDict and HowNet, as well as {adjectives, adverbs, verbs}, to conduct our optimization experiments. We used Figs. 4.2 and 4.3 to visualize the results of our experiments with varying values of C. By leveraging advanced machine learning techniques and evaluation metrics, our approach enables us to accurately classify sentiment in Chinese text data at scale, providing valuable insights into the underlying sentiment of large-scale text data sources in various domains.

Our results demonstrate that fine-tuning the regularization parameter C of SVM^{perf} slightly improves the performance of our approach in terms of both F1 and accuracy. Interestingly, we found that the two selected strategies achieved the best accuracy at different values of C. Specifically, the combination of IARDict and HowNet achieved the best accuracy when C = 10, while {adjectives, adverbs, verbs} performed best when C = 20. By leveraging advanced machine learning techniques and evaluation metrics, our approach enables us to accurately classify sentiment in Chinese text data at scale, providing valuable insights into the underlying sentiment of large-scale text data sources in various domains.

Our experiments have demonstrated that both the lexicon-based method and part-of-speech-based method can achieve excellent results, with a total classification accuracy of over 90%. In this subsection, we discuss the reasons for the effectiveness of our approach.

Firstly, the vector representations of words learned by word2vec enable us to extract deep semantic relationships between words, which are crucial for accurate sentiment classification. This allows our approach to capture the nuances and complexities of language use in Chinese text data at scale.

Secondly, as a new training algorithm for linear classification SVM based on SVM^{light}, SVM^{perf} is much more precise and faster than SVM^{light} for large-scale data sets. This allows us to effectively classify sentiment in large-scale text data sources, enhancing our understanding of the factors that contribute to sentiment expression in Chinese text data.

By leveraging advanced techniques and algorithms, our approach achieves encouraging performance in sentiment classification. By accurately classifying sentiment in Chinese text data at scale, our approach provides valuable insights into the underlying sentiment of large-scale text data sources in various domains, enhancing our understanding of the factors that contribute to sentiment expression in Chinese text data.

4.1.4 Conclusion

Unlike most conventional methods for sentiment classification that focus on simple lexical or syntactic features, our research focuses on the deep semantic relationships between words. In this section, we utilized advanced tools, including word2vec and SVM^{perf}, to classify Chinese comment texts. To conduct our experiments, we crawled tens of thousands of Chinese comments on clothing products to create a dataset. We first used word2vec to cluster similar features and found that it was also effective for Chinese text clustering. Our proposed approach achieved over 90% accuracy in sentiment classification, whether using the lexicon-based or part-of-speech-based feature selection method, demonstrating the effectiveness of our approach.

However, our research is not without limitations. One challenge we face is how to apply high-dimensional feature vectors to SVM^{perf}'s training file format, as the current dimension of vectors learned by word2vec is set to 1. Additionally, the two feature selection methods we used may not have been sufficient to identify all sentiment features in each sentence. Therefore, in future work, we plan to explore extracting structured information and composition units from every sentence to improve the accuracy of our approach. By leveraging advanced techniques and algorithms, our approach enables us to accurately classify sentiment in Chinese text data at scale, providing valuable insights into the underlying sentiment of large-scale text data sources in various domains.

4.2 Contextual Heterogeneous Feature Fusion Framework for Audio Sentiment Analysis

4.2.1 Introduction

Audio sentiment analysis is a challenging task that has gained significant attention from researchers. Emotion-related features, including statistic and spectrum features, have been developed to classify audio data accurately. However, combining these heterogeneous features effectively is challenging, as they reflect different aspects of audio data. To address this issue, we propose a novel heterogeneous feature fusion framework that consists of two stages: context-independent feature extraction and context-dependent representation learning. Our approach leverages the Residual Convolutional Model with Spatial Attention (RCMSA) and the Contextual Heterogeneous Feature Fusion Model (CHFFM) to integrate spectrum and statistical features. Our experiments on two public sentiment analysis datasets, MOSI and MOUD.

4.2.2 Proposed Method

Our heterogeneous feature fusion framework, as depicted in Fig. 4.4, involves two stages: context-independent feature extraction and context-dependent representation learning. In the first stage, we utilize the Residual Convolutional Model with Spatial Attention (RCMSA) to extract discriminative spectrum features from mel-spectrograms. In the second stage, we introduce the Contextual Heterogeneous Feature Fusion Model (CHFFM), which effectively integrates spectrum and statistical features, leveraging contextual information between utterances to achieve better audio sentiment analysis results. Additionally, we propose a Contextual Single Feature Model (CSFM) to evaluate the performance of each type of feature.

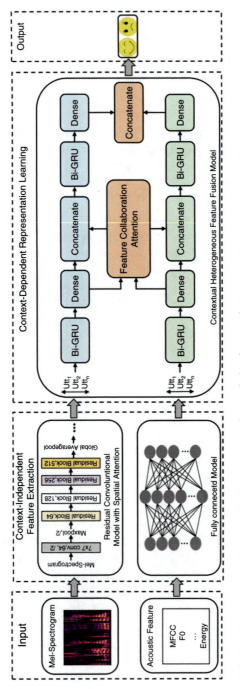

Fig. 4.4 Overall architecture of the proposed heterogeneous feature fusion framework

Context-Independent Feature Extraction

In this section, we focus on extracting both spectrum and statistical features from each individual utterance, without taking into account any contextual information between utterances.

1. **Spectrum Feature Extraction**: First, we process all audio files to 16000 framerate. Then, we apply 1024 length Fast Fourier Transform (FFT) windows on the windowed audio samples, using a hop length of 512 for the Short Term Fourier Transform (STFT). Finally, we map the spectrogram to the mel-scale to obtain the mel-spectrogram. The mel-spectrogram is a visual representation of the signal frequency, which is generated from the audio signal through STFT.

   We propose the Residual Convolutional Model with Spatial Attention (RCMSA) to extract discriminative spectrum features from mel-spectrograms, inspired by the Residual Convolutional Network with Self-Attention Model. The RCMSA model consists of four residual blocks, where the structure of each residual block is shown in Fig. 4.5. We employ the residual connection to maintain the original structure of the input data, which has been shown to be effective in previous works. Furthermore, we introduce spatial attention to the RCMSA model, which enables the model to import emotion-salient information from mel-spectrograms and provide discriminative feature learning results for the spectrum feature extraction process. To obtain the mel-spectrogram representation M, we input the mel-spectrograms into a 7×7 convolution layer and a max pooling layer before passing it through the residual blocks:

$$\text{Attention}(M) = \text{Softmax}(\text{Tanh}(\text{Conv}(M)))M \tag{4.1}$$

   To represent each mel-spectrogram as a spectrum feature vector, we apply a 1×1 convolution layer (Conv) followed by a global average pooling layer. This process reduces the dimensionality of the mel-spectrogram and extracts the most important information from it, resulting in a 512-dimension spectrum feature vector.

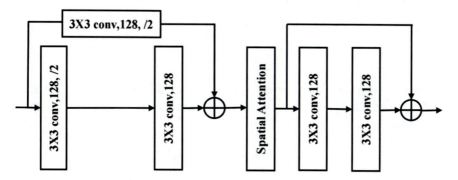

Fig. 4.5 The structure of the residual block with 128 filters

2. **Statistical Feature Extraction**: To extract statistical features from audio signals, we utilize openSMILE tool with the emobase2010 configuration file. This configuration file includes several low-level descriptors, such as MFCC, voice intensity, and pitch. Using this configuration file, we obtain 1,582 statistical features from each utterance, which is consistent with the features used in the INTERSPEECH 2010 Paralinguistic Challenge. However, to reduce the dimensionality difference with spectrum features, we utilize two fully connected layers to map the dimension of the statistical features from 1,582 to 500.

Context-Dependent Representation Learning

To leverage the contextual information between utterances, we divide the heterogeneous features extracted. This enables us to capture the relationship between different utterances and make full use of contextual information to achieve better audio sentiment analysis results. To effectively integrate spectrum and statistical features, we propose the Contextual Heterogeneous Feature Fusion Model (CHFFM). Additionally, we introduce the Contextual Single Feature Model (CSFM) to evaluate the performance of each type of feature and use it as a baseline to demonstrate the significance of heterogeneous feature fusion.

1. **Contextual Heterogeneous Feature Fusion Model**: The Contextual Heterogeneous Feature Fusion Model (CHFFM) is represented in Fig. 4.4, it takes two types of heterogeneous features as input: spectrum features extracted from mel-spectrograms, and statistical features extracted from audio signals. We begin by feeding these features into a Bi-GRU layer to learn low-level context information between utterances. Next, we utilize a fully connected layer to map the representations of these heterogeneous features to the same dimension. To maximize the benefits of each type of feature and ensure sufficient interaction between them, we introduce feature collaboration attention on the outputs of the fully connected layers, D_m and D_s. This attention mechanism computes a pair of matching matrices, M_m and M_s, over the two representations, taking into account the heterogeneous features information.

$$M_m = D_m D_m^T \tag{4.2}$$

$$M_s = D_s D_s^T \tag{4.3}$$

After computing the matching matrices, M_m and M_s, we compute the utterance-level attention matrices of the heterogeneous features, resulting in outputs O_m and O_s.

$$N_m = \text{Softmax}(\text{Tanh}(M_m)) \tag{4.4}$$

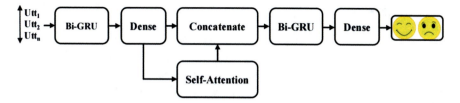

Fig. 4.6 The structure of the Contextual Single Feature Model

$$N_s = \text{Softmax}(\text{Tanh}(M_s)) \tag{4.5}$$

$$O_m = N_m D_s \tag{4.6}$$

$$O_s = N_s D_m \tag{4.7}$$

To fully capture the nature of each feature and obtain more comprehensive sentiment information, we concatenate the input and output of the feature collaboration attention.

$$C_m = \text{Concat}[D_m, O_m] \tag{4.8}$$

$$C_s = \text{Concat}[D_s, O_s] \tag{4.9}$$

To leverage the information contained in both spectrum and statistical features, we utilize another Bi-GRU layer followed by a fully connected layer to learn high-level contextual information from C_m and C_s. This process enables us to capture the complex relationships between different features and make full use of contextual information to improve the accuracy of sentiment classification. Finally, we concatenate the outputs of these fully connected layers to obtain the final classification results.

2. **Contextual Single Feature Model**: To assess the individual performance of each type of feature and demonstrate the importance of heterogeneous feature fusion, we introduce the Contextual Single Feature Model (CSFM). As depicted in Fig. 4.6, we utilize a Bi-GRU layer followed by a fully connected layer to learn low-level contextual information, similar to the CHFFM. However, instead of feature collaboration attention, we use self-attention [10] to capture important sentiment information. Next, we concatenate the input and output of the self-attention and pass them through another Bi-GRU layer to learn high-level contextual information. Finally, we leverage a fully connected layer to obtain the final sentiment classification results. By utilizing this approach, we can assess the individual contributions of spectrum and statistical features and demonstrate the significance of heterogeneous feature fusion.

4.2.3   Experiments

Datasets

We perform experiments on two publicly available sentiment analysis datasets, MOSI [11] and MOUD [12]. The MOSI dataset consists of 93 videos from 89 unique speakers, collected from YouTube movie reviews. The label of each utterance falls within a continuous range of -3 to $+3$, where -3 represents highly negative and 3 represents highly positive sentiments. We categorize labels in the range of $[-3,0)$ as negative and labels in the range of $[0,3]$ as positive. Considering the independence of the speakers, we split the dataset into 52 videos (1284 utterances) for the training set, 10 videos (229 utterances) for the validation set, and 31 videos (686 utterances) for the test set. The MOUD dataset comprises product review videos in Spanish, with each video containing multiple segments labeled as positive, negative, or neutral sentiment. We exclude the neutral sentiment label, resulting in a dataset of 59 videos (322 utterances) for the train and validation set, and 20 videos (116 utterances) for the test set. Our dataset splitting strategy aligns with previous works [13].

Baseline Models

To demonstrate the effectiveness of our proposed method, we conduct a comparative analysis with state-of-the-art models that utilize either spectrum features or statistical features. The models we compare our approach with are as follows:

1. **SC-LSTM** [14]: The Simple Contextual LSTM (SC-LSTM) model is a straight-forward adaptation of the contextual LSTM, comprising of unidirectional LSTM cells.
2. **BC-LSTM** [14]: The Bidirectional Contextual LSTM (BC-LSTM) model is designed to perform context-dependent sentiment analysis. This model primarily comprises of bidirectional LSTM cells, enabling it to capture contextual information in both forward and backward directions. Currently, it is considered the state-of-the-art model for sentiment analysis on the MOUD dataset [13].
3. **MU-SA** [15]: The Multiutterance-Self Attention (MU-SA) model is a state-of-the-art approach for sentiment analysis on the MOSI dataset. This model comprises of Bi-GRU cells and self-attention, which allows it to capture contextual information across multiple utterances and leverage it to improve the accuracy of sentiment classification.

Experimental Setup

Our model comprises of low-level Bi-GRU layers with 200 neurons and high-level Bi-GRU layers with 100 neurons. We utilize ReLU activation function in the dense layers and Softmax activation function in the output layer. To prevent overfitting, we apply a dropout rate of 0.7 to the Bi-GRU layers and dense layers for regularization. During the model training, we set the batch size to 32 and utilize Adam optimizer with cross-entropy loss function. We evaluate the performance of our method using Accuracy and F1-score as evaluation metrics. To ensure the reliability of the results, we perform five runs with different random seeds and consider the average of these runs as the final experiment results.

Experimental Results

Table 4.7 presents the results of our experiments on the MOSI dataset. The results show that our proposed CHFFM model outperforms all baseline models in terms of accuracy, F1-score, and other evaluation metrics, achieving a 2.16% improvement in accuracy over the best baseline result. Moreover, our proposed CSFM model achieves the best single feature accuracy, with 58.43% and 56.62% accuracy using spectrum features and statistical features, respectively. However, compared to the CSFM model, the CHFFM model is more efficient and achieves higher accuracy and F1-score, outperforming by 1.31% in accuracy and 1.72% in F1-score. This is mainly because the CHFFM effectively integrates heterogeneous features, enabling it to capture more comprehensive sentiment information and extract richer emotional characteristics.

Table 4.7 The experimental results for audio sentiment analysis on the MOSI dataset. M: spectrum features. S: statistical features

Model	Features	Accuracy	F1-score
SC-LSTM	S	52.71	53.96
SC-LSTM	M	56.56	52.74
BC-LSTM	S	57.35	54.56
BC-LSTM	M	55.31	48.96
MU-SA	S	57.58	53.40
MU-SA	M	56.36	52.01
CSFM(ours)	S	58.43	53.27
CSFM(ours)	M	56.62	48.29
CHFFM(ours)	S + M	**59.74**	**54.99**

Table 4.8 The experimental results for audio sentiment analysis on the MOUD dataset. M: spectrum features. S: statistical features

Model	Features	Accuracy	F1-score
SC-LSTM	S	61.55	59.42
SC-LSTM	M	65.69	61.7
BC-LSTM	S	58.62	52.7
BC-LSTM	M	64.31	56.04
MU-SA	S	62.93	56.93
MU-SA	M	67.24	61.95
CSFM(ours)	S	63.1	56.52
CSFM(ours)	M	67.41	66.35
CHFFM(ours)	S + M	71.38	67.89

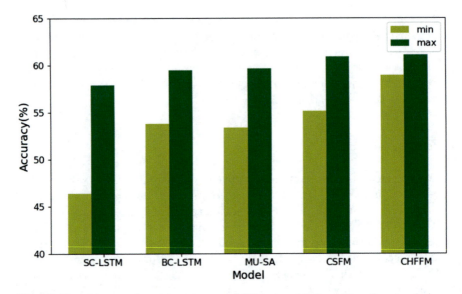

Fig. 4.7 The minimum and maximum accuracy of different models for audio sentiment analysis on the MOSI dataset

Table 4.8 presents the results of our experiments on the MOUD dataset. Our proposed CHFFM model significantly outperforms state-of-the-art methods in terms of accuracy and F1-score. Compared to the MU-SA model, the CHFFM achieves a relative improvement of 4.14% in accuracy and 5.94% in F1-score. Furthermore, compared to the BC-LSTM model, the CHFFM improves accuracy by 7.07%. Our proposed CSFM model also delivers excellent performance, achieving the best single feature accuracy and improving the best single feature F1-score from 61.95% to 66.35%. By comparing the results of the CHFFM and the CSFM, we observe that the CHFFM outperforms the CSFM by 3.97% in accuracy and 1.54% in F1-score. These results demonstrate the importance of integrating heterogeneous features for accurate audio sentiment analysis.

By combining the experimental results on the MOSI and MOUD datasets, we can confirm that our method outperforms current state-of-the-art models in audio sentiment analysis. The comparison between the CHFFM and CSFM experimental results also demonstrates that incorporating heterogeneous features can comprehensively learn emotional information and improve the performance of audio sentiment analysis. In particular, Fig. 4.7 shows the minimum and maximum accuracy of all models on the MOSI dataset. The CHFFM model demonstrates superior performance in both minimum and maximum accuracy, verifying the effectiveness of our model. Furthermore, the smallest difference between the minimum and maximum accuracy of the CHFFM model indicates high robustness of our model.

4.2.4   Conclusion

This section presents a framework for audio sentiment analysis that combines different types of features. The framework has two main stages: first, we use the Residual Convolutional Model with Spatial Attention (RCMSA) to extract important spectral features from mel-spectrograms in a context-independent way. Next, we use the Contextual Heterogeneous Feature Fusion Model (CHFFM) to combine the extracted spectral features with statistical features in a context-dependent manner. Our experimental results demonstrate that our framework outperforms previous state-of-the-art methods on both the MOSI and MOUD datasets. In the future, we plan to extend our approach to analyze sentiment in multiple modes and explore new ways to merge different types of features from different modes.

Image sentiment analysis based on coattentive multitask convolutional neural network.

4.2.5   Introduction

The goal of Facial Expression Recognition (FER) is to identify the emotional state conveyed by a facial image. FER is becoming increasingly important in a variety of areas, including human-computer interaction (HCI), video analysis, and social communication. The prototypical facial expressions that FER seeks to recognize include anger, disgust, fear, happiness, sadness, and surprise. These expressions are distinguished by subtle changes in facial muscle movements. However, FER is also impacted by objective factors such as lighting and variations in facial pose.

Over the past few years, FER researchers have focused on improving the performance of FER in both lab-controlled and real-world settings. These algorithms range from traditional hand-crafted features (such as Gabor filters [16], Histogram of Gradients [17], and Local Binary Patterns [18]) to deep neural networks (including Deep Convolutional Neural Networks [19] and Generative Adversarial Networks [20]). Several recent studies have proposed new techniques to improve the accuracy of FER. For example, Ding et al. [21] developed a two-stage training algorithm that fine-tunes the face network and improves expression recognition performance. Cai et al. [22] introduced a novel constraint function, called island loss, to learn discriminative features among different expressions. Wang et al. [23] designed a region attention network and used ensemble learning to improve generalizability and robustness, achieving state-of-the-art results on multiple public databases.

Therefore, we propose a novel end-to-end Coattentive Multitask Convolutional Neural Network (CMCNN), which integrates the Channel Coattention Module (CCAM) and the Spatial Coattention Module (SCAM). The CCAM captures the interdependencies of different channels between FER and FLD tasks to generate channel coattention scores. Meanwhile, the SCAM utilizes max- and average-pooling operations to formulate spatial coattention scores.

We conduct extensive experiments on four widely used benchmark facial expression databases, including RAF, SFEW2, CK+, and Oulu-CASIA. The results demonstrate that our approach outperforms both single-task and multitask baselines. Our proposed CMCNN model achieves superior performance, demonstrating the effectiveness and generalizability of multitask learning for FER.

4.2.6   Methodology

Our proposed approach for FER involves an end-to-end coattentive multitask convolutional neural network. In this section, we provide a detailed explanation of our approach. Firstly, we introduce the overall architecture of our model. Then, we focus on the two key components of our model: the channel coattention module and the spatial coattention module. These modules are designed to capture the interdependencies among different channels and spatial locations of the FER and FLD tasks. Finally, we describe the multitask loss function that we use to optimize the network. Our approach is designed to improve the overall performance of FER by leveraging the complementary information provided by FLD.

Coattentive Multitask Convolutional Neural Network

To ensure the scalability of our multitask learning approach, we use two identical deep convolutional neural networks (DCNNs) to perform the FER and FLD tasks separately. Each DCNN consists of six convolutional components, each of which comprises a convolutional layer, a batch normalization layer, a ReLU activation layer, and an optional max pooling layer. The architecture of each component is detailed in Table 4.9.

In our multitask learning approach, we refer to the outputs of the last convolutional component for the FLD and FER tasks as L_{in} and E_{in}, respectively. The inputs for the next convolutional component for the FLD and FER tasks are referred to as L_{out} and E_{out}, respectively. This naming convention helps us to clearly distinguish between the input and output of each component in the DCNNs for the FLD and FER tasks.

Table 4.9 BaseDCNN architecture for single tasks, FLD, and FER

Layer type	1 Conv	2 MP	3 Conv	4 MP	5 Conv	6 Conv	7 MP	8 Conv	9 Conv
Kernel	3	2	3	2	3	3	2	3	3
Output	64	–	96	–	128	128	–	256	256
Stride	1	2	1	2	1	1	2	1	1
Pad	1	0	1	0	1	1	0	1	1

The overall architecture of our Coattentive Multitask Convolutional Neural Network (CMCNN) is depicted in the upper part of Fig. 4.8, while the lower part provides a detailed description of the Channel Coattention Module (CCAM) on the left and the Spatial Coattention Module (SCAM) on the right. In our approach, we use two identical DCNNs for the FLD and FER tasks, with L_{in} and E_{in} representing the outputs of the last convolutional component for FLD and FER tasks, respectively, and L_{out} and E_{out} representing the inputs for the next convolutional component for FLD and FER tasks. We introduce a channel and spatial coattentive module (Co-ATT) between two adjacent convolutional components to capture interdependencies between the FLD and FER tasks.

The Co-ATT is composed of two modules: the CCAM and the SCAM. The CCAM generates channel coattention scores, L_{out}^{c} and E_{out}^{c}, by capturing the interdependencies of different channels between the FLD and FER tasks. The SCAM formulates spatial coattention scores, L_{out}^{s} and E_{out}^{s}, by combining max- and average-pooling operations. The final outputs of the Co-ATT are generated by reweighting the outputs of the CCAM and the SCAM. It is essential to note that we exchange the outputs of the CCAM to decrease feature redundancy and increase feature sharing between the FLD and FER tasks. We provide a detailed explanation of the CCAM and SCAM in the following two sections.

$$\begin{bmatrix} L_{out} & - \\ - & E_{out} \end{bmatrix} = \begin{bmatrix} \alpha_1 & \alpha_2 \\ \beta_1 & \beta_2 \end{bmatrix} \times \begin{bmatrix} E_{out}^{c} & L_{out}^{c} \\ L_{out}^{s} & E_{out}^{s} \end{bmatrix} \tag{4.10}$$

where α_1, α_2, β_1, and β_2 are trainable parameters.

To classify the extracted facial features from the base DCNN, we use fully connected layers as classifiers. In our approach, we adopt a 2-layer fully connected network for the FLD task and a 3-layer fully connected network for the FER task. The use of batch normalization is imperative to prevent overfitting. In our network, we use the ReLU activation function for all layers except the last one. The fully connected layers serve as a crucial component of our approach, allowing us to generate predictions for the FER and FLD tasks based on the extracted features from the DCNNs.

Coattentive Multitask Convolutional Neural Network

In convolutional neural networks, the output of the convolutional components usually contains multiple independent channel features that contribute differently to the final results. Therefore, it is crucial to apply different weights to different channels before sharing features, which is precisely what our Channel Coattention Module (CCAM) is designed to accomplish. For ease of reference in the following sections, we use C, H, and W to represent the number of channels, height, and width of feature maps, respectively. By weighting different channels appropriately, the

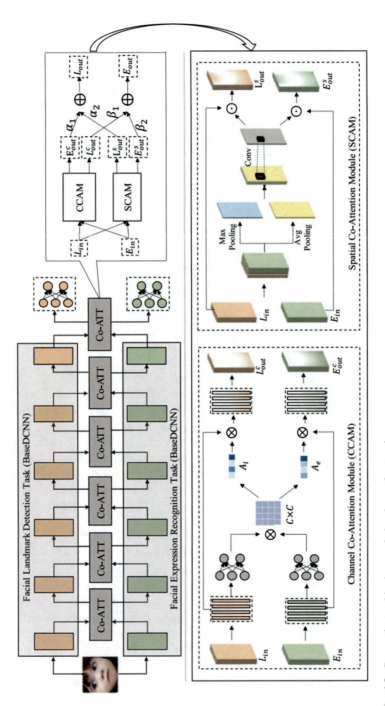

Fig. 4.8 Coattentive multitask convolutional neural network

CCAM helps to reduce feature redundancy and increase feature sharing between the FLD and FER tasks, ultimately improving the overall accuracy of our approach.

To enable our Channel Coattention Module (CCAM) to appropriately weight different channels in the FLD and FER tasks, we first use two independent two-layer fully connected networks to achieve feature nonlinear transformation. Specifically, we take inputs $L_{in} \in R^{C \times H \times W}$ from the FLD task and $E_{in} \in R^{C \times H \times W}$ from the FER task and apply the two-layer fully connected network separately to each input. This approach helps us avoid the overhead of high-dimensional calculations while transforming the input features into a more informative representation:

$$F_l = f\left(\widehat{L}_{in}; W_l\right) \tag{4.11}$$

$$F_e = f\left(\widehat{E}_{in}; W_e\right) \tag{4.12}$$

To achieve feature nonlinear transformation for the FLD and FER tasks, we first flatten the inputs L_{in} and E_{in} along the channel axis, resulting in $\widehat{L}_{in} \in R^{C \times HW}$ and $\widehat{E}_{in} \in R^{C \times HW}$, respectively. We then apply a multilayer perceptron (MLP) with trainable weights, W_l and W_e, to each flattened input. This MLP is implemented by a nonlinear activation function $f(\cdot)$, which helps to capture complex relationships between the input features and their representations. As a result of this transformation, we obtain two feature vectors, F_l and $F_e \in R^{C \times D}$, where $D = \delta \times H \times W$, $0 < \delta < 1$.

Building on the work of Vaswani et al. [10], we create the attention matrix by calculating the resemblance between distinct channel features:

$$S = \frac{F_e F_l^T}{\sqrt{D}} \in R^{C \times C} \tag{4.13}$$

The attention weight between the ith channel in the FER and jth channel in the FLD is represented by S_{ij}.

To normalize the attention scores, we employ the softmax function. Normalizing along different axes results in attention scores with different task preferences. In particular, for a channel in the FER task, we normalize S along the first C axis and evaluate the attention score with all channels in the FLD task:

$$S_{ij}^e = \frac{\exp\left(S_{ij}\right)}{\sum_{k=1}^{C} \exp\left(S_{kj}\right)}; \quad A_e = \sum_{j=1}^{C} S_{\cdot j}^e \tag{4.14}$$

where $S_{ij}^e \in R^{C \times 1}$ is the jth column in matrix S.

Likewise, if a channel belongs to the FLD task, we normalize S along the second C axis and calculate the attention score with all channels in the FER task:

$$S_{ij}^l = \frac{\exp(S_{ij})}{\sum_{k=1}^{C} \exp(S_{kj})}; \quad A_l = \sum_{j=1}^{C} S_{\cdot j}^l \qquad (4.15)$$

where $S_{ij}^l \in R^{1 \times C}$ is the ith column in matrix S.

Finally, we incorporate the channel attentions A_e and A_l into their respective original inputs, resulting in the output of CCAM:

$$E_{out}{}^c = A_e \odot E_{in} \qquad (4.16)$$

$$L_{out}{}^c = A_l \odot L_{in} \qquad (4.17)$$

where $\odot$ means element-size multiply operation. $E_{out}{}^c \in R^{C \times H \times W}$ and $L_{out}{}^c \in R^{C \times H \times W}$.

Spatial Coattention Module

In contrast to CCAM, SCAM emphasizes the interspatial relationship between features. To capture spatial information, we apply both max-pooling and average-pooling operations along the channel axis, which has been demonstrated to be effective in enhancing local detail information [24].

To begin with, we merge L_{in} and E_{in} along the channel axis and create two feature maps via max-pooling and average-pooling operations:

$$F_{max} = \text{MaxPool}[L_{in}; E_{in}] \qquad (4.18)$$

$$F_{avg} = \text{AvgPool}[L_{in}; E_{in}] \qquad (4.19)$$

where F_{max} and $F_{avg} \in R^{1 \times H \times W}$.

Next, we combine F_{max} and F_{avg} along the channel axis and apply a standard convolutional transformation to generate a shared attention map:

$$A_s = \sigma[F_{max}; F_{avg}] * W_s \qquad (4.20)$$

Here, $*$ denotes the convolution operation using kernel parameter $W_s \in R^{1 \times 7 \times 7}$. The sigmoid activation function σ is also applied.

Lastly, we apply A_s into the L_{in} and E_{in} and get the outputs of SCAM:

$$E_{out}{}^s = A_s \odot E_{in} \qquad (4.21)$$

$$L_{out}{}^s = A_s \odot L_{in} \qquad (4.22)$$

where $E_{out}{}^s$ and $L_{out}{}^s \in R^{C \times H \times W}$.

Multitask Loss

FLD is a regression task, while FER is a classification task. As a result, we apply distinct loss functions to supervise their training. Specifically, we utilize wing loss [25] as the supervision for the FLD task:

$$loss_L = \begin{cases} \omega \ln(1+X/\epsilon) & \text{if } X < \omega; \\ X - M & \text{otherwise} \end{cases} \qquad (4.23)$$

The wing loss function is defined as $X = |y_l - \widehat{y}_l|$, where ω and ϵ are two hyperparameters. ω specifies the range of the nonlinear segment as $(-\omega , \omega)$, while ϵ controls the curvature of the nonlinear region. We set $\omega = 10$ and $\epsilon = 2$, consistent with the work of Feng et al [25] The constant M is computed as $M = \omega - \omega \ln(1 + X/\epsilon)$.

For the task FER, we use cross entropy as the supervision:

$$loss_E = - [y_e \log \widehat{y}_e + (1 - y_e) \log(1 - \widehat{y}_e)] \qquad (4.24)$$

where y_e and $\widehat{y}_e$ are the true values and predictive values, respectively.

Then, the total optimization objective of our approach is:

$$loss = loss_E + \lambda \cdot loss_L \qquad (4.25)$$

where λ is a hyperparameter for controlling the weight of $loss_L$.

Methodology

We evaluate the effectiveness of our method through extensive experiments on four benchmark expression databases, consisting of two lab-controlled databases (CK+ [26, 27] and Oulu-CASIA [28]) and two real-world databases (RAF [29] and SFEW2 [30]). In addition to comparing various single-task and multitask baselines, we conduct three sets of transfer experiments to demonstrate the generalizability and robustness of our method in different data scenarios. Furthermore, we analyze the performance of our approach in terms of feature visualization, time cost (Table 4.18), and parameter analysis. Finally, we conduct a detailed ablation study to investigate the contributions of different modules in our approach. To provide a comprehensive understanding of our experiments, we provide a detailed introduction to our experimental settings in this section.

Benchmark Databases

1. **RAF**: RAF [29] is a real-world facial image database that contains 29,672 images collected from diverse search engines. The annotations in this database include

both basic and compound expressions, and we only utilize the images with basic expression annotations. Ultimately, we select 12,771 images for training and 3068 images for validation.

2. **SFEW2**: SFEW 2 [30] is the most extensively used benchmark database for recognizing facial expressions in the wild. It contains 1766 images, with 958 images for training, 436 images for validation, and 372 images for testing. Each image is labeled with one of seven expression categories, which includes neutral and the six basic expressions. The expression labels for the training and validation sets are provided, while the labels for the testing set are withheld by the challenge organizers. Hence, we report the results on the validation sets.

3. **CK+**: CK+ [26, 27] is a database that consists of 327 videos obtained from 118 subjects. These video sequences capture the facial expressions from neutral to the peak expression. Similar to Cai et al. [22], we select the last three frames of each sequence that correspond to the provided expression label. As a result, our experiments involve 981 images (7 expressions, including contempt and the six basic expressions).

4. **Oulu-CASIA**: The Oulu-CASIA database [28] consists of 2880 videos, each capturing one of the six basic expressions from 80 subjects. Similar to prior research conducted on the Oulu-CASIA database [22], we only utilize the 480 videos collected by the VIS system under normal indoor lighting conditions for our experiments. As with CK+, we select the last three frames of each video as the peak frames of the labeled expression. Therefore, the Oulu-CASIA database comprises 1440 images for our experiments.

Multitask baselines

In this study, we compare our method to three traditional multitask baselines. To ensure a fair comparison, we employ a consistent convolutional neural network for all methods, as presented in Table 4.9. In particular, we include a batch normalization layer and a ReLU activation layer after each convolutional layer. We use the terms "Conv" and "MP" to denote Convolution and MaxPooling Operation, respectively.

1. **HPS**: Hard Parameter Sharing (HPS) is a straightforward and intuitive multitask approach that involves sharing the bottom layer parameters and separating the top layer parameters.

2. **CSN**: Cross-Stitch Network (CSN) [31] is a technique that involves soft sharing and utilizes cross-stitch units to capture various split architectures of different tasks. Each task has its own independent, but identical, network architecture.

3. **PS-MCNN**: PS-MCNN [32] is a soft sharing approach that involves an extra shared network to facilitate the interaction between various task-specific networks.

Data Preprocessing

1. **Facial Landmark Annotations**: Table 4.10 illustrates that the different databases in our study have varying landmark annotations, and some do not have any annotations at all. To ensure consistency, we require a standardized label for the landmarks across all databases. While facial landmark detection is not our primary objective, we can utilize an established detection tool to fulfill our requirements. Therefore, we opt to use the OpenFace2.0 toolbox [33] to obtain 68 facial landmarks for all databases. In our approach, facial landmarks serve two purposes: aligning faces and supervising the FLD subtask.
2. Face Detection and Alignment: Facial alignment is a critical preprocessing step in expression recognition. Although the RAF database provides aligned faces, the other three databases do not. Therefore, we employ Multitask Cascade Convolutional Neural Networks (MTCNN) [34] to detect the face in the SFEW2, CK+, and Oulu-CASIA databases. After successful face detection, we resize all facial images to 100×100. We apply a three-point constrained affine transformation for face alignment, utilizing the coordinates of the left eye, right eye, and the midpoint of the corners of the lips for alignment.
3. Data Augmentation: As the training data is relatively limited and we do not use any additional data, we apply online data augmentation methods to augment the training set. These methods include horizontal flipping and rotating images to the left or right at angles between -10 and $10°$. We restrict these operations to the training set only.

Experimental Details

This section provides a comprehensive overview of our experimental details, which encompasses our training/testing strategy, training specifics, hyperparameter selection, and evaluation metrics.

1. Training/Testing Strategy: Our method and all multitask baselines are trained and tested using static images. We follow the approach taken by Cai et al. [22] and

Table 4.10 Facial expression and facial landmarks annotations statistics of RAF, SFEW2, CK+, and Oulu-CASIA. "NA" means vacancy

#	RAF	SFEW2	CK+	Oulu-CASIA
	Happy	Happy	Happy	Happy
Types of expression	Neutral	Neutral	Contempt	
	Sad	Sad	Sad	Sad
	Disgust	Disgust	Disgust	Disgust
	Fear	Fear	Fear	Fear
	Surprise	Surprise	Surprise	Surprise
	Angry	Angry	Angry	Angry
Number of landmarks	5 or 37	NA	68	NA

utilize a tenfold cross-validation strategy for the CK+ and Oulu-CASIA databases. Each database is split into ten subsets, and the subjects in any two subsets are mutually exclusive. For each run, we use data from eight sets for training, and the remaining two subsets are reserved for validation and testing, respectively.

2. Training Details: We initiate the training of our model from scratch, utilizing Adam as the optimizer with an initial learning rate of 0.01. The learning rate decay strategy involves reducing the learning rate by a factor of ten every ten epochs. We also apply L2 normalization for model weights, with a weight decay of 0.005. The training process terminates when the model's performance on the validation set fails to improve for eight epochs. We select the best model to obtain the results on the test set.

3. Hypermeter Selection: To adjust the hyperparameters for all multitask learning baselines and our method, we adopt the grid search strategy. To ensure a fair comparison, we conduct each experiment with five random seeds, which include 1, 12, 123, 1234, and 12,345. We report the average performance across these seeds.

4.2.7 Experiments

Comparisons with Multitask Methods

In this section, we present a comparative analysis of CMCNN with three multitask learning methods and the single-task baseline, BaseDCNN. We report the results in Tables 4.11 and 4.12.

In this section, we conduct a comparative analysis of the multitask methods and single-task baseline for the FER task. We observe that all multitask methods outperform the single-task baseline on the four benchmark databases, demonstrating the feasibility and significance of including FLD as an auxiliary task for FER. However, it is worth noting that the performance relies on the quality of the landmark annotations, and more precise annotations could further enhance the results.

Next, we compare the performance of our method with three multitask baselines. Our results indicate that our method outperforms all baselines on all evaluation metrics. Notably, our approach exhibits significant improvement over "CSN", which shares a similar architecture to ours, indicating that our Co-ATT module plays a crucial role in advancing FER performance. In summary, our findings suggest that multitask learning methods are beneficial for FER, and our approach significantly enhances FER performance.

Table 4.11 (%) Results for FER on RAF and SFEW2 with different multitask learning methods "BaseDCNN (FER)" and "BaseDCNN (FLD)" are the single-task baselines for FER and FLD tasks, respectively

Method	RAF				SFEW2			
	ACC_{total}	ACC_{averag}	F1	NRMSE	ACC_{total}	ACC_{averag}	F1	NRMSE
BaseDCNN(FER)	82.73	73.19	74.83	–	32.98	30.81	29.82	–
BaseDCNN(FLD)	–	–	–	3.73	–	–	–	22.3
HPS	83.02	73.78	75.21	3.88	35.32	32.74	32.33	40.36
CSN	85.10	75.49	77.50	4.07	34.03	30.91	30.34	35.03
PS-MCNN	84.67	76.31	77.12	3.81	35.32	32.00	30.92	49.79
CMCNN	85.22	77.03	77.97	3.71	37.95	34.95	34.39	27.81

Table 4.12 (%) Results for FER on CK+ and Oulu-CASIA with different multitask learning methods "BaseDCNN (FER)" and "BaseDCNN (FLD)" are the single-task baselines for FER and FLD tasks, respectively

Method	CK+				Oulu-CASIA			
	ACC_{total}	ACC_{averag}	F1	NRMSE	ACC_{total}	ACC_{averag}	F1	NRMSE
BaseDCNN(FER)	92.75	94.10	93.64	–	83.46	83.46	83.46	–
BaseDCNN(FLD)	–	–	–	6.37	–	–	–	3.32
HPS	94.31	93.05	92.21	29.35	80.74	80.74	80.64	10.12
CSN	95.59	94.37	93.74	23.82	83.54	83.54	83.46	9.61
PS-MCNN	96.16	94.92	94.42	48.73	83.49	83.49	83.41	8.19
CMCNN	96.71	96.02	95.48	14.87	85.04	85.04	85.35	4.64

Comparisons with State-of-the-Arts

This section presents a comparative analysis between our experimental results and other state-of-the-art methods. Some of the state-of-the-art methods employ deeper CNN architectures and are pretrained on larger face databases. To ensure a fair comparison, similar to SCN [35] and RAN [23], we augment our CMCNN by incorporating ResNet18 [36] pretrained on the MS-Celeb-1 M face recognition dataset. We refer to this modified model as CMCNN-ResNet18 in the following sections. Additionally, we reproduce the experimental results with ResNet18 as the new single-task baseline.

1. **RAF**: Table 4.13 presents a comparison of our method with other state-of-the-art approaches on the RAF database. The left half of the table depicts the comparison results, while the right half displays the confusion matrix of the CMCNN evaluated on the RAF validation set. The confusion matrix indicates the classification performance of different expressions, such as happiness (Ha), contempt (Co), sadness (Sa), disgust (Di), fear (Fe), surprise (Su), and angry (An). We compare our approach with DLP-CNN [29], DSAN-RES-AGE [37], SCN [35], and RAN [23]. Although DLP-CNN [29] uses the same DCNN backbone as CMCNN, our approach outperforms it by almost 3%. While RAN [23] achieves the best performance for the overall accuracy (ACC_{total}), our new method, CMCNN-ResNet18, obtains a higher performance of 90.36%, surpassing RAN by 2%. DSAN-RESAGE [37] achieves the best performance for the average accuracy ($ACC_{average}$) by adopting a two-stage training strategy instead of an end-to-end model. However, our approach achieves better results and is an end-to-end model. Additionally, CMCNN-ResNet18 demonstrates significant improvement over ResNet18, which is consistent with the comparison between CMCNN and BaseDCNN.

   The right half of Table 4.13 displays the confusion matrix of CMCNN evaluated on the RAF test set. We observe that the model achieves the highest performance for the "Ha" expression category, while the performance for "Di" and "Fe" categories is significantly lower than the other categories. This observation aligns with the distribution of data categories, indicating that the model's performance reflects the class distribution of the dataset.

2. **SFEW2**: Table 4.14 presents a comparative analysis of the performance of our method with other state-of-the-art approaches on the SFEW2 database. The right half of the table displays the confusion matrix of the CMCNN evaluated on the SFEW2 validation set, where each row corresponds to the ground truth expression category, and each column represents the predicted expression category. We compare our approach with AUDN [43], STM-ExpLet [44], IL-VGG [22], and SFEW best [45]. Similar to the RAF database, SFEW2 is also a real-world database with a small sample size. Furthermore, SFEW2 is a competition database that utilizes various competition strategies such as additional training data, multiple network fusion, and careful parameter adjustment to enhance its performance.

Table 4.13 (%) Comparisons with state-of-the-arts on the RAF database

Method	ACCtotal	ACCaverage		Ha	Ne	Sa	Di	Fe	Su	An
DLP-CNN [38]	–	74.20	Ha	**93.1**	3.78	1.13	0.62	0.14	0.73	0.56
DSAN-RES-AGE [39]	–	76.40	Co	3.41	**84.7**	6.65	2.12	0.0	2.71	0.47
SCN [40]	88.14	–	Sa	3.43	8.12	**83.8**	2.68	0.33	0.67	0.96
RAN [41]	88.55	–	Di	8.87	13.5	11.0	**54.4**	1.87	2.37	8.00
ResNet18 [42]	86.29	78.33	Fe	4.59	5.41	10.8	3.78	61.1	9.73	4.59
CMCNN	85.22	77.03	Su	2.74	5.11	2.67	0.73	2.92	84.1	1.7
CMCNN-ResNet18	90.36	82.26	An	5.06	4.81	2.22	5.56	1.73	2.47	78.2

Table 4.14 (%) Comparisons with state-of-the-arts on the SFEW2 database

Method	ACCtotal	ACCawerage		Ha	Ne	Sa	Di	Fe	Su	An
STM-ExpLet	–	31.73	Ha	59.7	10.6	11.1	1.39	1.39	1.67	14.2
IL-VGG	–	44.95	Co	4.52	49.5	20.7	3.10	6.90	9.52	5.71
SFEW best	–	52.50	Sa	7.32	22.54	31.8	5.35	6.76	16.9	9.30
SCN	54.19	–	Di	8.18	20.9	26.4	20.9	9.10	1.80	12.7
ResNet18	42.37	39.01	Fe	6.98	13.5	14.4	6.51	18.6	26.5	13.5
CMCNN	37.95	34.39	Su	12.7	13.1	10.4	1.15	10.0	38.5	14.2
CMCNN-ResNet18	45.78	41.56	An	21.6	13.6	13.9	7.50	3.70	14.1	25.6

Table 4.15 (%) Comparisons with state-of-the-arts on the CK+ database

Method	ACCtotal	ACCaverage		Ha	Co	Sa	Di	Fe	Su	An
MSR [46]	–	91.40	Ha	**99.4**	0.3	0.0	0.0	0.3	0.0	0.0
F-Bases [47]	–	94.81	Co	0.0	**91.9**	6.7	0.7	0.4	0.4	0.0
IL-VGG	–	91.64	Sa	0.0	2.5	**93.1**	0.0	0.0	0.0	3.3
IL-CNN	–	94.39	Di	0.3	0.3	0.0	**97.6**	0.0	0.3	1.4
ResNet18	96.93	96.54	Fe	0.0	0.0	0.0	0.0	**96.0**	4.0	0.0
CMCNN	96.71	96.02	Su	0.0	1.4	0.2	0.2	0.1	**98.2**	0.0
CMCNN-ResNet18	98.33	97.52	An	0.0	1.0	2.8	1.2	0.0	0.0	**94.9**

Table 4.16 (%) Comparisons with state-of-the-arts on the Oulu-CASIA database

Method	ACCtotal	ACCaverage		Ha	Sa	Di	Fe	Su	An
FN2EN	87.71	87.71	Ha	92.4	2.4	0.6	3.4	0.2	1.0
STM-ExpLet	74.59	74.59	Sa	1.8	80.2	2.9	2.1	0.0	13.1
IL-CNN	77.29	77.29	Di	1.2	2.2	79.8	2.9	0.9	13.1
IL-VGG	84.58	84.58	Fe	6.5	3.2	2.3	83.8	1.7	2.6
ResNet18	84.92	84.92	Su	0.1	0.0	0.3	2.1	96.7	0.8
CMCNN	85.04	85.04	An	0.2	7.2	10.3	2.2	0.1	79.8
CMCNN-ResNet18	87.32	87.32							

3. **CK**: In Table 4.15, we present a comparative analysis of our method with MSR [46], F-Bases [47], IL-VGG [22], and IL-CNN [22] on the CK+ database. The left half of the table displays the comparison results, while the right half shows the confusion matrix of the CMCNN evaluated on the CK+ testing set. The confusion matrix depicts the classification performance of different expressions, such as happiness (Ha), contempt (Co), sadness (Sa), disgust (Di), fear (Fe), surprise (Su), and angry (An).

4. **Oulu-CASIA**: Table 4.16 presents a comparative analysis of our approach with other state-of-the-art methods on the Oulu-CASIA database. The left half of the table shows the comparison results, while the right half displays the confusion matrix of the CMCNN evaluated on the Oulu-CASIA testing set. The confusion matrix indicates the classification performance of different expressions, such as happiness (Ha), sadness (Sa), disgust (Di), fear (Fe), surprise (Su), and angry (An).

We compare our approach with FN2EN, STM-ExpLet, IL-CNN, and IL-VGG. The Oulu-CASIA database is category-balanced, so ACC_{total} is equal to $ACC_{average}$. Our approach achieves comparable performance on the Oulu-CASIA database. It is worth noting that the state-of-the-art method, FN2EN Ding et al., utilizes additional face recognition databases and adopts a two-stage training strategy, which is not an end-to-end network.

Transfer Validation

To ensure that our method can be applied to a wide range of situations, we conducted transfer validation experiments between real-world and lab-controlled databases using three sets: "Real & Lab," "Real & Real," and "Lab & Lab." Each set of experiments included mutual validation, where we trained the model with one database and tested it on another database, and vice versa. For instance, in the "Real & Lab" experiments, we used RAF as the training set and CK+ as the testing set, and then we reversed the roles. We repeated each experiment five times and reported the average performance on the "Test Set." As different databases have different emotional categories (as detailed in Table 4.10), we removed the "Neural" category in RAF and SFEW databases and the "Contempt" category in the CK+ database. Consequently, we conducted six classification experiments in this section, covering all possible scenarios.

Table 4.17 presents the results of our experiments. The terms "Real" and "Lab" represent the characteristics of the datasets. RAF and SFEW2 are real-world databases, while CK+ and Oulu ("Oulu-CASIA") are lab-controlled databases. The number below each database name indicates the size of the dataset. The "BaseDCNN" is the single-task model used as a baseline. We conducted each experiment five times and reported the average values on the "Test Set." Our method achieved significant improvements in all data scenarios, indicating that it has better generalizability and learns more transferable features. When comparing the results in "Real & Lab," we found that the model trained on real-world databases had better transferability than the one trained on lab-controlled databases. Moreover, the results in "Real & Real" showed that transfer learning for real-world facial expression recognition remains challenging. Finally, when comparing the results in "Lab & Lab", we noticed that the performance of "from Oulu to CK+" was significantly better than that of "from CK+ to Oulu." This is because the Oulu dataset has more

Table 4.17 (%) Transfer validation on different train sets and test sets

Type	Train Set	Valid Set	Test Set	Model	ACCaverae	F1
Real	RAF	RAF	CK+	BaseDCNN	56.90	54.63
&	(12,771)	(3068)	(981)	CMCNN	**63.72**	**64.16**
Lab	CK+	CK+	RAF	BaseDCNN	27.10	24.00
	(784)	(197)	(15,839)	CMCNN	**29.86**	**26.68**
Real	RAF	RAF	SFEW2	BaseDCNN	32.87	29.85
&	(12,771)	(3068)	(1394)	CMCNN	**33.95**	**30.27**
Real	SFEW2	SFEW2	RAF	BaseDCNN	24.29	20.25
	(958)	(436)	(15,839)	CMCNN	**26.41**	**21.13**
Lab	CK+	CK+	Oulu	BaseDCNN	40.04	36.04
&	(784)	(197)	(1440)	CMCNN	**46.54**	**42.96**
Lab	Oulu	Oulu	CK+	BaseDCNN	67.04	62.12
	(1152)	(288)	(981)	CMCNN	**70.21**	**66.39**

samples and larger face variations, making it more diverse and representative of real-world scenarios.

Feature Visualization

We conducted a visualization study on the features learned by our approach and the baseline using t-SNE [48]. Figure 4.9 shows the results of this study, which includes a total of 981 samples from the CK+ database and 15,339 samples (12,271 training data and 3068 testing data) from the RAF database. The learned features are clustered according to the 7 expressions, with the results for CK+ displayed in the upper part and those for RAF in the lower part of Fig. 4.9. Compared to the BaseDCNN, our approach produced more compact clusters with significantly fewer outliers. This demonstrates that our method can learn more significant features, which decreases the intraclass feature differences and increases the variations of interclass features. The visualization study provides further evidence that our approach has better generalizability and can learn more transferable features than the baseline. The results are best viewed in color.

Time Cost Analysis

To fully assess the generalizability of our method, we conduct three sets of transfer validation experiments between real-world and lab-controlled databases, namely "Real & Lab," "Real & Real," and "Lab & Lab." In each set of experiments, we perform mutual validation experiments wherein we use one database for training and another for testing. For instance, in the "Real & Lab" experiments, we train our model on the RAF database and test it on the CK+ database. Subsequently, we train our model on the CK+ database and test it on the RAF database. We report the average performance of five trials on the "Test Set." It is worth noting that different databases have different emotional categories, as illustrated in Table 4.10. Therefore, we remove the "Neutral" category in RAF and SFEW databases and the "Contempt" category in the CK+ database to ensure consistency in the evaluation metrics.

4.2.8   Conclusion

We presented a novel approach for facial expression recognition by introducing facial landmarks detection as an auxiliary task. Our approach is an end-to-end coattentive multitask convolutional neural network that comprises the CCAM and the SCAM. We conducted extensive experimental evaluations on multiple databases, including RAF, SFEW2, CK+, and Oulu-CASIA, demonstrating the effectiveness and robustness of our method. We also conducted multiple sets of transfer

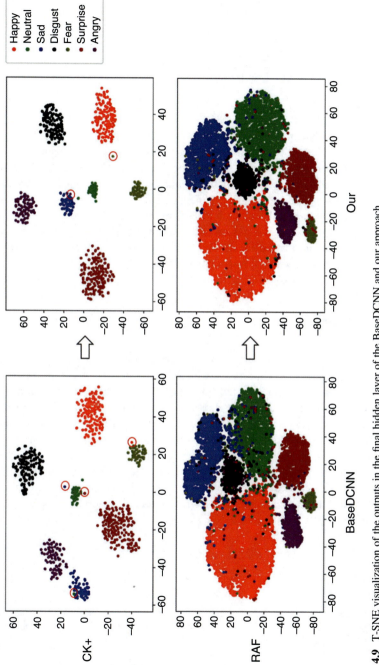

Fig. 4.9 T-SNE visualization of the outputs in the final hidden layer of the BaseDCNN and our approach

Table 4.18 Time cost comparison between our method and baselines on the RAF database during the inference stage. Batch = 32

Method	Time (ms)/Batch
SCN	47.82
RAN	134.19
CBAM	42.67
BaseDCNN	25.43
ResNet18	38.17
CMCNN	42.18
CMCNN-ResNet18	73.53

validation experiments to validate the method's transferability. Our detailed model analysis revealed that the CCAM played a more significant role in our approach, and the SCAM further improved our model's performance. Our method offers new possibilities for facial expression recognition and can be easily applied to other tasks.

In future work, we plan to investigate the interactivity between face-related tasks, including facial landmarks detection, facial expression recognition, and facial action units detection. We will also focus on addressing the occluded facial problem and enhancing the model's robustness. Our approach has significant potential for advancing the field of facial expression recognition.

4.3   Summary

This chapter delves into sentiment analysis in various domains, including text, audio, and facial expressions. It explores advanced techniques such as word2vec and SVM^{perf} for classifying Chinese comment texts based on deep semantic relationships between words. This chapter also presents a framework for audio sentiment analysis, combining spectral and statistical features, which outperforms existing methods. Additionally, a novel approach for facial expression recognition is introduced, incorporating facial landmarks detection as an auxiliary task. This chapter concludes by highlighting the effectiveness and potential of these approaches in advancing sentiment analysis across different domains. Overall, this chapter provides valuable insights and sets the stage for further research and practical applications in the field of sentiment analysis.

References

1. Zhang D, Lee WS (2006) Extracting key-substring-group features for text classification. Proceedings of the 12th Association for Computing Machinery Special Interest Group on Knowledge Discovery and Data Mining international conference on Knowledge discovery and data mining, 474–483
2. Tomas M, Kai C, Greg C (2013) Efficient estimation of word representations in vector space. Proceedings of the International Conference on Learning Representations, 1–12

3. Mikolov T, Yih WT, Zweig G (2013) Linguistic regularities in continuous space word representations. Proceedings of the 2013 conference of the North American chapter of the association for computational linguistics: Human language technologies, 746–751

4. Zhai Z, Liu B, Xu H, et al (2010) Grouping product features using semisupervised learning with soft-constraints. Proceedings of the 23rd international conference on computational linguistics, 1272–1280

5. Zhai Z, Liu B, Xu H, et al (2011) Clustering product features for opinion mining [C]. Proceedings of the fourth Association for Computing Machinery international conference on Web search and data mining, 347–354

6. Bing L (2012) Sentiment analysis and opinion mining. Synth Lect Hum Lang Technol 5(1): 1–167

7. Liu B, Zhang L (2012) A survey of opinion mining and sentiment analysis[J]. Mining text data, 415–463

8. Pang B, Lee L, Vaithyanathan S (2002) Thumbs up? sentiment classification using machine learning techniques. Proceedings of the Association for Computational Linguistics-02 conference on Empirical methods in natural language processing, 79–86

9. Tang H, Tan S, Cheng X (2009) A survey on sentiment detection of reviews. Expert Syst Appl 36(7):10760–10773

10. Vaswani A, Shazeer N, Parmar N, et al (2017) Attention is all you need. Proceedings of the 31st International Conference on Neural Information Processing Systems, 5998–6008

11. Zadeh A, Zellers R, Pincus E (2016) Multimodal sentiment intensity analysis in videos: facial gestures and verbal messages. IEEE Intell Syst 31(6):82–88

12. Pérez-Rosas V, Mihalcea R, Morency LP (2013) Utterance-level multimodal sentiment analysis. Proceedings of the 51st Annual Meeting of the Association for Computational Linguistics, 1: 973–982

13. Zadeh A, Liang PP, Poria S, et al (2018) Multi-attention recurrent network for human communication comprehension. Proceedings of the 32nd Association for the Advancement of Artificial Intelligence Conference on Artificial Intelligence, 5642–5649

14. Poria S, Cambria E, Hazarika D, et al (2017) Context-dependent sentiment analysis in usergenerated videos. Proceedings of the 55th annual meeting of the association for computational linguistics, 1: 873–883

15. Ghosal D, Akhtar MS, Chauhan D, et al (2018) Contextual inter-modal attention for multimodal sentiment analysis. Proceedings of the 2018 Conference on Empirical Methods in Natural Language Processing, 3454–3466

16. Liu C, Wechsler H (2002) Gabor feature based classification using the enhanced fisher linear discriminant model for face recognition[J]. IEEE Trans Image Process 11(4):467–476

17. Dalal N, Triggs B (2005) Histograms of oriented gradients for human detection. Proceedings of the 2005 Institute of Electrical and Electronics Engineering Computer Society Conference on Computer Vision and Pattern Recognition, 1: 886–893

18. Shan C, Gong S, McOwan PW (2009) Facial expression recognition based on local binary patterns: a comprehensive study. Image Vis Comput 27(6):803–816

19. Krizhevsky A, Sutskever I, Hinton GE (2012) Imagenet classification with deep convolutional neural networks. Commun ACM 60:84–90

20. Goodfellow IJ, Abadie JP, Mirza M, et al (2014) Generative adversarial nets[C]. Proceedings of the 27th International Conference on Neural Information Processing Systems, 2672–2680

21. Ding H; Zhou SK; Chellappa R (2017) Facenet2expnet: regularizing a deep face recognition net for expression recognition[C]. Proceedings of the 12th Institute of Electrical and Electronics Engineering International Conference on Automatic Face & Gesture Recognition, 118–126

22. Cai J, Meng Z, Khan AS, et al (2018) Island loss for learning discriminative features in facial expression recognition. Proceedings of the 13th Institute of Electrical and Electronics Engineering International Conference on Automatic Face & Gesture Recognition, 302–309

23. Wang K, Peng X, Yang J et al (2020) Region attention networks for pose and occlusion robust facial expression recognition. IEEE Trans Image Process 29:4057–4069

24. Woo S, Park J, Lee JY, et al (2018) CBAM: convolutional block attention module. Proceedings of the European Conference on Computer Vision, 3–19

25. Feng ZH, Kittler J, Awais M, et al (2018) Wing loss for robust facial landmark localisation with convolutional neural networks. Proceedings of the Institute of Electrical and Electronics Engineering Conference on Computer Vision and Pattern Recognition, 2235–2245

26. Lucey P, Cohn JF, Kanade T, et al (2010) The extended Cohn-Kanade dataset (CK+): a complete dataset for action unit and emotion-specified expression. Proceedings of the Institute of Electrical and Electronics Engineering Computer Society Conference on Computer Vision and Pattern Recognition Workshops, 94–101

27. Kanade T, Cohn JF, Tian Y. (2000) Comprehensive database for facial expression analysis. Proceedings of the 4th Institute of Electrical and Electronics Engineering International Conference on Automatic Face and Gesture Recognition, 46–53

28. Zhao G, Huang X, Taini M et al (2011) Facial expression recognition from near-infrared videos. Image Vis Comput 29(9):607–619

29. Li S, Deng W (2018) Reliable crowdsourcing and deep locality-preserving learning for unconstrained facial expression recognition[J]. IEEE Trans Image Process 28(1):356–370

30. Dhall A, Goecke R, Lucey S, et al (2011) Static facial expressions in tough conditions: data, evaluation protocol and benchmark. Proceedings of the 2011 Institute of Electrical and Electronics Engineering International Conference on Computer Vision Workshops, 2106–2112

31. Misra I, Shrivastava A, Gupta A, et al (2016) Cross-stitch networks for multitask learning. Proceedings of the Institute of Electrical and Electronics Engineering Conference on Computer Vision and Pattern Recognition, 3994–4003

32. Cao J, Li Y, Zhang Z (2018) Partially shared multi-task convolutional neural network with local constraint for face attribute learning. Proceedings of the Institute of Electrical and Electronics Engineering Conference on Computer Vision and Pattern Recognition, 4290–4299

33. Baltrusaitis T, Zadeh A, Lim YC, et al (2018) OpenFace 2.0: Facial Behavior Analysis Toolkit [C]. Proceedings of the 13th Institute of Electrical and Electronics Engineering International Conference on Automatic Face & Gesture Recognition, 59–66

34. Zhang K, Zhang Z, Li Z et al (2016) Joint face detection and alignment using multitask cascaded convolutional networks. IEEE Sig Process Lett 23(10):1499–1503

35. Wang K, Peng X, Yang J, et al (2020) Suppressing uncertainties for large-scale facial expression recognition. Proceedings of the Institute of Electrical and Electronics Engineering Conference on Computer Vision and Pattern Recognition, 6897–6906

36. He K, Zhang X, Ren S, et al (2016) Deep residual learning for image recognition. Proceedings of the 2016 Institute of Electrical and Electronics Engineering Conference on Computer Vision and Pattern Recognition, 770–778

37. Fan Y, Li VOK, Lam JCK (2020) Facial expression recognition with deeply-supervised attention network[J]. IEEE Trans Affect Comput 13(2):1057–1071

38. Ngiam J, Khosla A, Kim M, et al (2011) Multimodal deep learning. Proceedings of the 28th International Conference on International Conference on Machine LearningJune, 689–696

39. Misra I, Shrivastava A, Gupta A, et al (2016) Cross-stitch networks for multi-task learning. Proceedings of the Institute of Electrical and Electronics Engineers conference on computer vision and pattern recognition, 3994–4003

40. Ma J, Zhao Z, Yi X, et al (2018) Modeling task relationships in multi-task learning with multi-gate mixture-of-experts. Proceedings of the 24th Association for Computing Machinery Special Interest Group on Knowledge Discovery and Data Mining, 1930–1939

41. Zadeh A, Liang PP, Mazumder N, et al (2018) Memory fusion network for multi-view sequential learning. Proceedings of the 32nd Association for the Advancement of Artificial Intelligence Conference on Artificial Intelligence, 5634–5641

42. Zhao X, Li H, Shen X, et al (2018) A modulation module for multi-task learning with applications in image retrieval. Proceedings of the European Conference on Computer Vision, 401–416

43. Liu M, Li S, Shan S, et al (2013) Au-aware deep networks for facial expression recognition. Proceedings of the 10th Institute of Electrical and Electronics Engineering International Conference and Workshops on Automatic Face and Gesture Recognition, 1–6

44. Liu M, Shan S, Wang R, et al (2014) Learning expressionlets on spatio-temporal manifold for dynamic facial expression recognition. Proceedings of the 2014 Institute of Electrical and Electronics Engineering Conference on Computer Vision and Pattern Recognition, 1749–1756

45. Kim BK, Roh J, Dong SY (2016) Hierarchical committee of deep convolutional neural networks for robust facial expression recognition[J]. Multimodal user. Interfaces 10(2): 173–189

46. Ptucha RW, Tsagkatakis G, Savakis AE (2011) Manifold based sparse representation for robust expression recognition without neutral subtraction. Proceedings of the 2011 Institute of Electrical and Electronics Engineering International Conference on Computer Vision Workshops, 2136–2143

47. Sariyanidi E, HaticeGunes AC (2017) Learning bases of activity for facial expression recognition. IEEE Trans Image Process 26(4):1965–1978

48. van der Maaten L, Hinton G (2008) Visualizing data using t-sne[J]. J Mach Learn Res 9(86): 2579–2605

Chapter 5
Cross-Modal Sentiment Analysis

Abstract This chapter explores multimodal sentiment analysis in social computing, highlighting the limitations of traditional text analysis and the importance of incorporating nonverbal modalities. It proposes an automatic MSA approach with MMix strategy, leveraging supervised and unsupervised data to accurately predict sentiment and reduce manual efforts. The results show state-of-the-art performance. The second section introduces a hierarchical model for acoustic analysis, combining frame-level and utterance-level features. It validates the model on IEMOCAP and MELD datasets. In the third section, this chapter presents CM-BERT, fine-tuning BERT with multimodal information. Experimental results on CMU-MOSI and CMU-MOSEI datasets show significant improvements. This chapter concludes by discussing the effectiveness of masked multimodal attention.

5.1 The Acoustic Visual Mixup Consistent (AV-MC) Framework

5.1.1 Introduction

Social media sentiment computing is a crucial research area in social computing applications. Traditional text sentiment analysis is not sufficient to predict users' sentiments accurately. As a result, recent research has shifted toward multimodal sentiment analysis, which uses streaming media resources that include text, acoustic, and visual modalities. However, manually collecting and annotating multimodal data is time-consuming, and understanding nonverbal modalities data is challenging.

To address these challenges, we proposed an automatic Multimodal Sentiment Analysis (MSA) approach with a Modality Mixup strategy (MMix) that emphasizes mining acoustic and visual implicit emotions. Our approach utilizes both annotated manually collected data (supervised data) and raw data from social media (unsupervised data). Additionally, the MMix strategy is considered an augmentation, which mixes the acoustic and visual modalities from different instances to generate potential emotion-bearing instances, improving the model's awareness of nonverbal modalities' implicit sentiment.

We conducted supervised and semisupervised experiments on social media resources and obtained state-of-the-art model performance in all metrics. The results indicate that our approach can accurately predict sentiment and minimize the time cost of manual statistics. Our proposed approach offers a promising solution to multimodal sentiment analysis, especially with respect to nonverbal modalities data.

5.1.2 Multimodal Sentiment Analysis (MSA) Background

With the increasing diversification of social media, multimodal sentiment analysis has become a popular research area in recent years. Figure 5.1 displays the four parts of the multimodal sentiment analysis (MSA) task, which are described below.

Multimodal Dataset Construction

The multimodal dataset is comprised of emotion-bearing videos collected from social media platforms. The videos are segmented based on a complete sentence, and only clips with clear faces, low ambient noise, and accurate text are preserved. After manual collection of the videos, instance annotations are conducted. The data is then split into train, validation, and test datasets according to a predetermined ratio. This final dataset serves as the foundation for the sentiment analysis task. Instances with manual emotion annotations are considered supervised data, while those lacking emotion labeling are considered unsupervised data.

Modality Feature Extraction

Modal feature extraction involves acquiring the raw information of each modality and transforming it into a symbolic feature that can be used in neural network calculations. The BERT pretraining model is used to extract character tokens from the original text, and corresponding sentence representations are obtained. For audio, the Librosa tool is employed to extract speech signal changes from each frame. For video, facial key points or emotion-related features such as behavior and action can be extracted.

Multimodal Fusion

Multimodal fusion involves combining information from various modalities for classification or regression tasks. There are two main approaches to multimodal fusion: model-independent and model-based approaches. Model-independent approaches do not rely on specific machine learning techniques for multimodal fusion and can be divided into early fusion, late fusion, and mixed fusion.

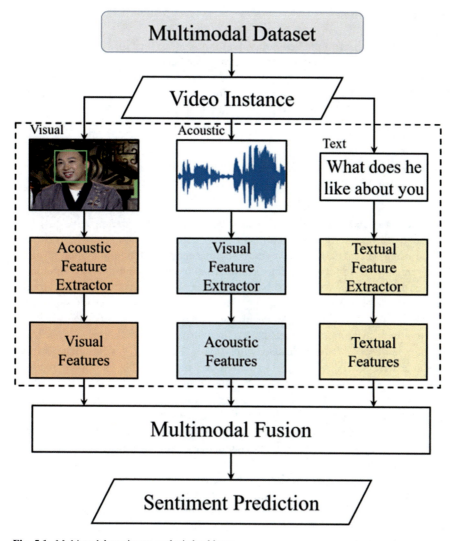

Fig. 5.1 Multimodal sentiment analysis backbone

On the other hand, model-based approaches rely on various machine learning techniques such as multiple kernel learning, probabilistic models, and neural networks for multimodal fusion. These approaches typically involve creating a model that can learn the relationships between the different modalities and use this information to improve the accuracy of the classification or regression task.

Sentiment Prediction

The process of sentiment prediction involves utilizing the final fusion representation to generate the predicted output. In regression prediction, the model forecasts the sentiment value within a specific range, such as $[-3, 3]$ in datasets like MOSI/MOSEI. On the other hand, in classification prediction, the model predicts the emotional state, such as happy or sad, in datasets like IEMOCAP.

5.1.3   *Automatic Sentiment Computing Approach with Modality Mixup Strategy*

Constructing datasets from the rich resources available in social media is just the initial step toward automating sentiment prediction. A prediction model must be trained using these datasets. However, nonverbal modality training is often insufficient, which led us to focus on acoustic and visual modalities during model training. Inspired by literature, we tried to balance the text by constructing various virtual acoustic and visual instances. To achieve this, we designed an automatic sentiment computing approach with a modality mixup strategy, as illustrated in Fig. 5.2.

The approach comprises three parts: single-task MSA backbone with MMix training, multitask MSA backbone with MMix$^{(*)}$ training, and multitask MSA backbone with MMix$^{(*)}$training. The dotted box represents the extra participation in the corresponding modality mixup training strategy. The approach is applied to two types of MSA task backbones: single-task MSA backbone and multitask MSA backbone. The single-task MSA backbone only requires multimodal manual annotations for supervised learning, while the multitask MSA framework requires text, acoustic, visual, and multimodal manual annotations for supervised learning.

Apart from the three different MSA task backbones, there are also three distinct modality mixup strategies (MMix, MMix$^{(*)}$, and MMix$^{(s*)}$) utilized between the single-task MSA backbone and multimodal MSA backbone. The MMix strategy is applied to the single-task supervised learning MSA backbone, MMix strategy is applied to the multitask supervised learning MSA backbone, and MMix$^{(s*)}$ strategy is applied to the multitask semisupervised learning MSA backbone.

Figure 5.3 depicts the different modality mixup strategies, these strategies enable us to balance the training of different modalities and improve the model's accuracy and performance.

Sentiment Prediction

The primary goal of the modality mixup strategy is to generate potential modality representations with corresponding annotations from the original modality encoder's output. To illustrate, let's consider two video instances from social media. As shown

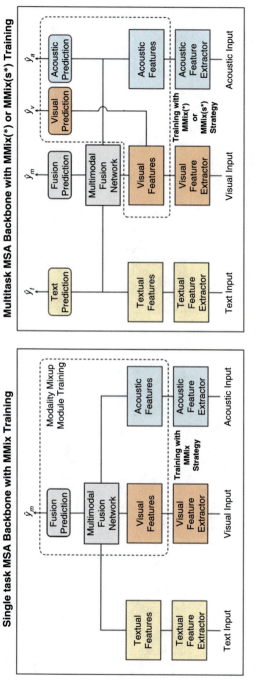

Fig. 5.2 Automatic sentiment computing approach with modality mixup strategy

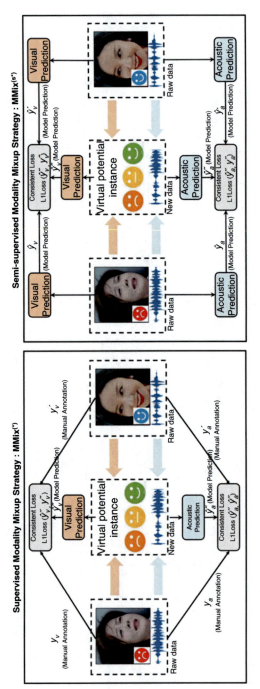

Fig. 5.3 Modality Mixup strategy

in Fig. 5.3, the MMix$^{(*)}$ situation refers to the training strategy under supervised data, while MMix$^{(s*)}$ refers to the training strategy under both supervised and unsupervised data. Both of these training strategies are performed on a multitask backbone, and the MMix scenario (modality mixup strategy on a single-task backbone) is omitted.

To generate a virtual nonverbal representation of intermediate emotion, we mix the acoustic and visual representation from a sad instance with the acoustic and visual representation from another happy instance. Similarly, we construct more virtual potential instances from the entire dataset. The set of instances with corresponding annotations, $\{(X_1, y_1), \ldots, (X_n, y_n)\}$, undergoes a random shuffle initially.

During training, we randomly select two instances, (X_i, y_i) and (X_j, y_j), from the shuffled set and mix their modality representations to generate a new instance, (X_{mix}, y_{mix}), which is then used to update the model parameters. This process continues until the end of the epoch. The modality mixup strategy helps to enhance the model's ability to learn and generalize across different modalities.

$$\{(X_1', y_1'), \ldots, (X_n', y_n')\} = \text{Shuffle}(\{(X_1, y_1), \ldots, (X_n, y_n)\}) \qquad (5.1)$$

Once the instances with corresponding annotations are shuffled, the mixup process is applied to the shuffled and original instance pairs using weighted averaging.

$$X_i^{''} = \lambda \cdot X_i + (1 - \lambda) \cdot X_i' \qquad (5.2)$$

$$y_i^{''} = \lambda \cdot y_i + (1 - \lambda) \cdot y_i' \qquad (5.3)$$

The modality mixup module is formulated using the definition that involves a random variable, λ, sampled from a Beta distribution within the range of $[0,1]$.

$$D_{mix} = \left\{ (X_1^{''}, y_1^{''}), \ldots, (X_n^{''}, y_n^{''}) \right\} = \text{MMix}\{(X_1, y_1), \ldots, (X_n, y_n)\} \qquad (5.4)$$

The virtual instances, D_{mix}, generated through the modality mixup strategy can serve as potential instances that facilitate representation learning.

The Training Process of Automatic Sentiment Computing Approach

The proposed Automatic Sentiment Computing Approach includes three model training scenarios.

One of the three model training scenarios in the proposed Automatic Sentiment Computing Approach is the single-task MSA backbone with supervised MMix strategy. In this scenario, the model utilizes supervised multimodal data with manual sentiment annotations for all modalities. The features of all three modalities are fused and classified by a classifier. Additionally, the acoustic and visual modalities both participate in the supervised modality mixup strategy training, where they are supervised by multimodal annotations.

Another model training scenario in the proposed Automatic Sentiment Computing Approach is the multitask MSA backbone with supervised MMix$^{(*)}$ strategy. In this scenario, the model utilizes supervised multimodal data with text, acoustic, visual, and multimodal manual sentiment annotations. The features of all three modalities are fused and classified by a classifier. Moreover, each individual modality independently predicts sentiment and is supervised by its own annotations, as shown in MMix$^{(*)}$ of Fig. 5.3. Additionally, both the acoustic and visual modalities participate in the multitask supervised modality mixup strategy, MMix$^{(*)}$, where they are supervised by their own acoustic and visual annotations separately.

The third model training scenario in the proposed Automatic Sentiment Computing Approach is the multitask MSA backbone with semisupervised MMixs strategy. In addition to the supervised data containing multiple modality annotations, the model incorporates unlabeled video clips from social media as unsupervised instances to participate in semisupervised training. The multitask training process is similar to the MMix strategy. However, the modality mixup strategy in this scenario differs from the previous ones. Since the unsupervised data lacks annotations, the prediction results of the two real data instances are used to supervise the generated mixed data, as shown in MMix$^{(s*)}$ of Fig. 5.3.

1. **Common Part**: The initial text, audio, and vision sequences are denoted as $I_k \in R^{T_k \times d_k(s*)}$, where T_k represents the length of the modality sequence, and d_k represents the initial feature dimension for modality $k(k \in \{t, a, v\})$.

    The unimodal encoders are then applied to the initial modality sequences, both for supervised and unsupervised instances, to learn intermodal representations. The unified unimodal encoders are formulated as follows:

$$F_m = S_k(I_k) \in R^{h_k} \tag{5.5}$$

    The formula above shows the unified unimodal encoders used in the proposed approach, where $k \in \{t, a, v\}$, h_k is the hidden dimension for modality k, and $S_u(\cdot)$ represents the unimodal encoder network.

    Specifically, for the textual modality, a one-layer feed-forward network is used as the encoder that transforms the first time step vector, which refers to the [CLS] token, into the textual representation. For the acoustic and visual modalities, a stacked bidirectional Long Short-Term Memory (LSTM) [1] followed by a one-layer feedforward network is utilized as the encoder.

    To obtain the final multimodal representation, we use the concatenation of the unimodal representation as a simple but effective fusion strategy.

$$F_m = \mathrm{Concat}([F_t; F_a; F_v]) \in R^{h_t + h_a + h_v} \tag{5.6}$$

    Once the unimodal and multimodal representations are obtained, different supervised training processes are used for different scenarios. For supervised instances, we utilize four independent classifiers, each consisting of a three-layer feed-forward network, for both the unimodal and multimodal sentiment predictions.

$$\widehat{y_k} = \text{Clf}_k(F_k) \in R \qquad (5.7)$$

The formula above shows the classifiers used in the proposed approach for both unimodal and multimodal sentiment prediction, where $k \in \{m, t, a, v\}$ and Clf refers to the predictive network used to convert the representation output into a regression value.

For each unimodal and multimodal task, L1 loss is used as supervision during training.

$$L_r{}^{(k)} = \frac{1}{N_s} \sum_{i=1}^{N_s} \left| \widehat{y_k}^{(i)} - y_k^{(i)} \right| \qquad (5.8)$$

For each $k \in \{m, t, a, v\}$, N_s represents the count of supervised instances. The final sentiment regression loss is formulated as a weighted average of the unimodal and multimodal tasks.

$$L_r = \sum_k a_k \cdot L_r{}^{(k)} \qquad (5.9)$$

The final sentiment regression loss is a weighted average of the unimodal and multimodal tasks, where aa_k, $k \in \{m, t, a, v\}$, is the hyperparameter that balances the contribution of each task.

In addition to the regression tasks for the supervised instances, mixup consistent tasks are performed for both supervised and unsupervised data on acoustic and visual modalities. However, the supervised and semisupervised approaches are handled differently.

2. **Supervised MMix$^{(*)}$**: The supervised modality mixup strategy is applied to the acoustic and visual representations, F_k, along with their corresponding manual annotations, y_k.

$$D_{\text{mix}}{}^{(k)} = \left\{ \left(F_k^{'(1)}, y_k^{'(1)} \right), \ldots, \left(F_k^{'(N_s)}, y_k^{'(N_s)} \right) \right\}$$
$$= \text{MMix}^{(*)} \left(\left\{ \left(F_k^{(1)}, y_k^{(1)} \right), \ldots, \left(F_k^{(N_s)}, y_k^{(N_s)} \right) \right\} \right) \qquad (5.10)$$

For $k \in \{a, v\}$, and N_s representing the total instance count for supervised data from the dataset, the supervised modality mixup strategy is applied. Using the generated acoustic and visual unimodal representations, the same acoustic and visual unimodal classifiers are utilized for both the mixed acoustic unimodal and mixed visual unimodal sentiment prediction.

$$\widehat{y}_k' = \text{Clf}_k \left(F_k' \right) \in R \tag{5.11}$$

The model prediction $\widehat{y}_k'$ is determined by the generated instance F_k'. Based on literature [2], the generated instances are considered as the interpolation between two original instances, and the prediction should be consistent with the weighted average of manual annotations of the original instances with the same weights, that is, $\widehat{y}_k'$. $L1$ Loss is still used for mixup consistent tasks.

$$L_{\text{mix}}^{(k)} = \frac{1}{N_s} \sum_{i=1}^{N_s} \left| \widehat{y}_k'^{(i)} - y_k'^{(i)} \right| \tag{5.12}$$

The acoustic and visual mixup tasks are combined by computing their weighted average to form the final consistent loss, where $k \in \{a, v\}$.

$$L_{\text{mix}} = \sum_k \beta_k \cdot L_{\text{mix}}^{(k)} \tag{5.13}$$

The hyperparameter β_k, where $k \in \{a, v\}$, is used to balance the impact of the acoustic and visual mixup tasks.

3. **Semisupervised MMix$^{(s*)}$**: The Semisupervised modality mixup strategy is applied to the acoustic and visual representation F_k and their corresponding prediction $\widehat{y}_k$

$$
\begin{aligned}
D_{\text{mix}}^{(k)} &= \left\{ \left(F_k'^{(1)}, \widehat{y}_k'^{(1)} \right), \ldots, \left(F_k'^{(N_s+N_u)}, \widehat{y}_k'^{(N_s+N_u)} \right) \right\} \\
&= \text{MMix}^{(s*)} \left(\left\{ \left(F_k^{(1)}, \widehat{y}_k^{(1)} \right), \ldots, \left(F_k^{(N_s+N_u)}, \widehat{y}_k^{(N_s+N_u)} \right) \right\} \right)
\end{aligned}
\tag{5.14}
$$

For $k \in \{a, v\}$, where N_s, N_u represent the total count of supervised data instances from the dataset and unsupervised data instances from social media platforms. We obtained the model predictions y_k'' by passing the mixed acoustic and visual instances F_k' through the same unimodal sentiment prediction network. Unlike the supervised approach, the newly generated instances are supervised using the weighted average of the predictions of the original instances. $L1\_Loss$ is still used for mixup consistent tasks.

$$L_{\text{mix}}^{(k)} = \frac{1}{N_s + N_u} \sum_{i=1}^{N_s+N_u} \left| \widehat{y}_k''^{(i)} - \widehat{y}_k'^{(i)} \right| \tag{5.15}$$

By taking into account both acoustic and visual mixup tasks, the final consistent loss is computed as the weighted average, where $k \in \{a, v\}$.

$$L_{\text{mix}} = \sum_k \beta_k \cdot L_{\text{mix}}^{(k)} \qquad\qquad (5.16)$$

The hyperparameter β_k is used to balance the relative contribution of acoustic and visual mixup tasks, where $k \in \{a, v\}$.

During semisupervised training, each training epoch consists of two iterations. In the first iteration, both supervised and unsupervised instances are used to update the model parameters with the help of the regression loss L_r and the mixup consistent loss L_{mix}. To ensure model convergence, a second iteration is performed on only the supervised instances using only the regression loss for supervision.

5.1.4   Experiments

Dataset

We conducted experiments on CH-SIMS [3], which is a Chinese MSA benchmark with fine-grained annotations of modality. The dataset comprises of 2281 refined video clips that were collected from various movies, TV serials, and variety shows with spontaneous expressions, occlusions, various head poses, and illuminations. Each sample was rated by human annotators with a sentiment score ranging from -1 (strongly negative) to 1 (strongly positive). We expanded the training instances using social media resources such as YOUTUBE, FACEBOOK, and BiliBili, and increased the supervised data to 4000 instances. Additionally, we relabeled the expanded supervised and unsupervised data to a total of 10,000 instances.

Feature Extraction

Different modal features were extracted as follows: For text modality, the pretrained bert-base-chinese model from BERT [4] was used to obtain effective textual features in the form of contextual word embeddings. The token sequences were either padded or truncated to a fixed length of 50, resulting in a word vector sequence in 768 dimensions as the final textual feature. Acoustic features were extracted using the OpenSMILE [5] backend at a sampling rate of 16,000 Hz. Specifically, the 25-dimensional eGeMAPS [5] Low Level Descriptors (LLD) features were used, and the resulting acoustic features were padded or truncated to a sequence length of 925. For visual modality, images were first extracted using FFmpeg at a rate of 25 frames per second. The TalkNet [6] method was then employed to detect the speaker's face among all faces in a single image. Instances with over 25% missing images or those where the ASD failed were discarded. The OpenFace [5] backend was used to extract facial features such as 68 facial landmarks, 17 facial action units, head pose, head orientation, and eye gaze direction. Finally, the 177-dimensional frame-level visual features were padded or truncated to a sequence length of 232.

Metrics

To evaluate the model performance, we followed previous work and used both classification and regression metrics. The traditional classification metrics such as binary classification accuracy (Acc2) and F1 score (F1_Score) were used to assess the accuracy of basic sentiment polarity prediction, that is, positive or negative classification. Additionally, Acc2_weak was used to evaluate the model performance for instances with weak emotions labeled in the range of $[-0.4, 0.4]$. For fine-grained prediction evaluation, we used traditional regression metrics like Mean Absolute Error (MAE) and the Pearson Correlation (Corr), which were formulated in Eq. (5.17).

$$\text{MAE}\left(Y, \widehat{Y}\right) = \frac{1}{n}\sum_{i=1}^{n} |y_i - \widehat{y}_i|$$

$$\text{Corr}\left(Y, \widehat{Y}\right) = \frac{\sum_{i=1}^{n}(y_i - \bar{y})(\widehat{y}_i - \bar{y})}{\sqrt{\sum_{i=1}^{n}(y_i - \bar{y})^2}\sqrt{\sum_{i=1}^{n}(\widehat{y}_i - \bar{y})^2}} \tag{5.17}$$

The average of the ground truth is denoted as $\bar{y} = \frac{1}{n}\sum_{i=1}^{n} y_i$. Additionally, we used R^2 to compare the model performance against the trivial solution, which involved predicting the average result on the test set. This was formulated in Eq. (5.18).

$$R^2\left(Y, \widehat{Y}\right) = \frac{\sum_{i=1}^{n}(y_i - \bar{y})^2 - \sum_{i=1}^{n}(\widehat{y}_i - \bar{y})^2}{\sum_{i=1}^{n}(\bar{y} - y_i)^2} \tag{5.18}$$

The average of the ground truth is represented by $\bar{y} = \frac{1}{n}\sum_{i=1}^{n} y_i$. In all the metrics mentioned above, higher values indicate better model performance, except for MAE, where lower values indicate better model performance.

Baselines

The benchmark models can be divided into two categories: those that use unified multimodal annotations for supervision and those that utilize unimodal annotations to guide unimodal representation learning. The former includes traditional MSA models such as the Late Fusion Deep Neural Network (LF DNN) [7], the Tensor Fusion Network (TFN) [8], the Low-rank Multimodal Fusion (LMF) [9], the Memory Fusion Network (MFN) [10], the Graph Memory Fusion Network (Graph MFN) [11], the Multimodal Transformer (MulT) [12], the Multimodal Adaptation Gate for

Bert Network (Bert MAG) [13], the Modality-Invariant and Modality-Specific Representations Network (MISA) [14], the Multimodal InfoMax Network (MMIM) [15], and the Self-Supervised multitask Learning Network (Self MM) [16]. The latter category includes models such as the Multitask Tensor Fusion Network (MTFN), the Multitask Late Fusion Deep Neural Network (MLF_DNN), and the Multitask Low-rank Multimodal Fusion (MLMF).

Supervised Sentiment Analysis

1. **Single Task MSA Backbone Experiment**: In our initial experiments, we used the single task MSA backbone, with all models supervised using the same multimodal annotations (as shown in the left of Fig. 5.2). We evaluated the model performances on our constructed dataset and found that while some models achieved good results on the positive and negative classification (Acc2) metrics, all benchmark models showed relatively low performance on the Acc2_weak metric. This suggests that existing models are only capable of identifying instances with strong sentiment expression and struggle to predict instances with weak sentiment expression. Moreover, some benchmark models exhibited extremely poor performance on the MAE metric, indicating that most existing models find it challenging to achieve accurate prediction of the correct sentiment value. However, the proposed model demonstrated an improvement in all metrics, particularly in the Acc2_weak metric (Table 5.1).

   In summary, our results indicate that the current MSA benchmarks struggle with fine-grained sentiment intensity prediction, despite performing well in basic sentiment polarity classification.

2. **Multitask MSA Backbone Experiment**: We also conducted experiments using the multitask MSA backbone, with all models supervised using both multimodal

Table 5.1 Model performances for single task Multimodal Sentiment Analysis model on CH-SIMS dataset. The best results are highlighted in bold

Models	Acc2($\uparrow$)	F1 score ($\uparrow$)	Acc2 weak ($\uparrow$)	Corr($\uparrow$)	R squre($\uparrow$)	MAE($\downarrow$)
LF_DNN	73.95	73.84	69.13	52.19	20.84	0.381
TFN	76.51	76.31	66.27	66.65	35.90	0.323
LMF	77.05	77.02	69.34	63.75	40.64	0.343
MFN	75.27	75.24	66.46	60.60	32.26	0.355
Graph_MFN	73.98	73.62	69.82	49.71	13.78	0.396
MulT	79.50	79.59	69.61	70.32	47.15	0.317
Bert_MAG	79.79	79.78	71.87	69.09	43.08	0.334
MISA	80.53	80.63	70.50	72.49	50.59	0.314
MMIM	80.95	80.97	72.28	70.65	43.81	0.316
Self MM	79.01	78.89	71.87	64.03	29.36	0.335
MMix	81.83 (1.11%)	81.85 (1.09%)	73.57 (1.78%)	73.17 (0.94%)	50.62 (0.06%)	0.311 (0.96%)

Table 5.2 Model performances for Traditional Multimodal Sentiment Analysis model on CH-SIMS v2.0 dataset. Models with (∗) are trained on multitasking. The best results are highlighted in bold

Models	Acc2($\uparrow$)	F1 score ($\uparrow$)	Acc2 weak ($\uparrow$)	Corr($\uparrow$)	R squre($\uparrow$)	MAE($\downarrow$)
MLF_DNN (∗)	78.40	78.44	71.59	65.80	39.34	0
MTFN(∗)	80.26	80.33	71.07	70.54	46.07	0.318
MLMF(∗)	79.92	79.72	69.88	71.37	47.53	0
MMix(∗)	**82.50** **(2.79%)**	**82.55** **(2.76%)**	**74.54** **(4.12%)**	**73.17** **(2.52%)**	**50.65** **(6.56%)**	**0.297** **(1.66%)**

annotations and unimodal annotations (as shown on the right of Fig. 5.2). Table 5.2 presents the model performances on our constructed dataset. We also compared the results of the two different class benchmarks and found that models with a simple late fusion backbone, which took advantage of the unimodal annotations, achieved competitive performances with state-of-the-art models. This further demonstrates the effectiveness of using unimodal annotations for fine-grained sentiment intensity prediction.

The proposed MMix approach demonstrated significant performance improvement, particularly in the Acc2_weak metric, when using the supervised settings of multitask training. The modality mixup strategy proved to be highly effective for unimodal annotation supervision in multitask training, as it allows each modality's representation to be independently enhanced, while also balancing the conflict between modalities to the fullest extent. The results of our experiments serve to validate the effectiveness of the modality mixup strategy for discriminating instances with weak emotion.

3. **Supplementary Verification Experiment**: We conducted additional experiments to verify the generalization of the proposed MMix approach, including unimodal and cross-modality sentiment analysis experiments. Unimodal sentiment analysis was performed using both unimodal and multimodal annotations, with the left side utilizing unimodal annotation for supervision and evaluation, while the right side used multimodal annotation for supervision and evaluation. Additionally, we separately applied the modality mixup strategy to the acoustic and visual modalities for presentation enhancement training. For all unimodal experiments, we utilized the same LSTM [1] based model for sentiment intensity regression, and the experimental results are presented in Table 5.3. The relatively poor performance observed in the acoustic unimodal task highlights the challenge of emotion-bearing acoustic feature extraction, which requires further research. Moreover, the performance gap observed between unimodal and multimodal annotations confirms the assumption that multimodal annotation can be misleading for unimodal representation learning. In longitudinal comparison, we observed partial improvement in the model's performance after using the proposed MMix strategy, as shown in Table 5.4. Furthermore, applying the MMix

Table 5.3 Model performances for unimodal sentiment analysis with modality mixup strategy. Models with (†) are trained with modality mixup strategy

Models	Unimodal labels				Multimodal labels			
	Acc2 ($\uparrow$)	Acc2 weak ($\uparrow$)	Corr ($\uparrow$)	MAE ($\downarrow$)	Acc2 ($\uparrow$)	Acc2 weak ($\uparrow$)	Corr ($\uparrow$)	MAE ($\downarrow$)
T	78.72	67.12	75.45	0.252	75.21	62.49	59.49	0.371
A	57.16	55.61	13.48	0.425	56.48	54.41	13.72	0.491
A(†)	58.01	58.51	14.31	0.424	57.35	58.11	15.33	0.489
V	78.72	76.04	57.70	0.314	73.11	68.79	48.54	0.401
V(†)	78.97	76.01	58.71	0.311	73.25	68.85	50.53	0.399

Table 5.4 Model performances for cross-modality sentiment analysis with modality mixup strategy. Models with (†) are trained with modality mixup strategy. Models with (∗) are trained on multitasking

Models	Acc2($\uparrow$)	Acc2 weak($\uparrow$)	Corr($\uparrow$)	MAE($\downarrow$)
A&V	73.11	68.38	50.94	0.394
A&V(†)	73.47	68.61	52.23	0.394
A&V(†∗)	73.69	68.79	52.48	0.386

Table 5.5 Model performance for Semisupervised Sentiment Analysis. Models with (∗) are trained on multitasking, (s∗ represents the

Models	Acc2($\uparrow$)	Acc2 weak($\uparrow$)	Corr($\uparrow$)	MAE($\downarrow$)
MMix$^{(*)}$	82.50	74.54	50.65	0.297
MMix$^{(s*)}$	83.46	74.54	57.37	0.286

strategy in cross-modal sentiment analysis of acoustic and visual modalities also resulted in performance improvement (Table. 5.5).

Semisupervised Sentiment Analysis

1. **Quantitative analysis**: MMix is a modality mixup strategy applied in sentiment analysis, which constructs virtual instances that can only simulate potential acoustic and visual instances, but cannot fully construct real instances, thus limiting the model's performance. To improve the model's performance, we can leverage unlabeled video clips on social media for semisupervised learning, which are cheaper and available in large quantities, and can be fully utilized by the model. In our experiments, we collected 10,000 unlabeled data from social media platforms and combined them with existing supervised data for semisupervised learning. The experimental results showed that unsupervised data can improve the model's performance, especially in the R^2 regression metric. The proposed MMix strategy significantly improved after adding additional unsupervised data for semisupervised training. Therefore, real unsupervised

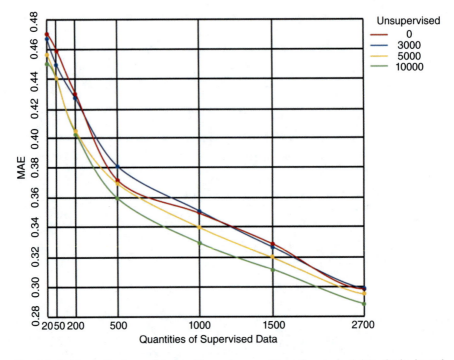

Fig. 5.4 Semisupervised experiments with varying quantities of supervised data, the horizontal axis represents the quantity of supervised data, the vertical axis represents MAE performance, and different colours represent the quantity of fixed and different unsupervised data. Fixed the quantity of unsupervised data, changed the quantity of supervised data, and conducted semisupervised experiments

data is more helpful for emotion prediction than the virtual samples generated by the modality mixup strategy. This is because real unsupervised data can provide more emotional expressions and variations, which help the model learn emotional features better. However, virtual samples can only simulate potential emotional expressions and variations, and cannot cover the diversity and complexity of real data, thus having limited impact on improving the model's performance. Therefore, leveraging real unsupervised data for semisupervised learning is an effective method to improve the performance of sentiment prediction.

2. **Qualitative analysis**: As shown in Fig. 5.4, this is an experiment to investigate the role of unsupervised data in the training process. We conducted a series of semisupervised experiments and compared the performance of the model under different ratios of supervised to unsupervised data. The experimental results showed that when the amount of supervised data is close to that of unsupervised data, the performance of semisupervised learning is poor. However, when the magnitude of unsupervised data is more than 2 times or more than that of supervised data, semisupervised training can be more effective.

In addition, the constructed dataset contains rich emotion-bearing nonverbal cues that alleviate the text-predominant phenomenon. Moreover, the independent unimodal annotation along with the unsupervised instances further provides a potential solution for exploring the nonverbal context. The proposed MMix strategy, which learns to be aware of nonverbal context through modality mixup, can be regarded as an initial attempt at the challenge of making full use of nonverbal behaviors. In the future, different feature combinations may be developed to further improve the performance of sentiment analysis.

5.1.5   Conclusion

Social media is a treasure trove of audio and video modal information, and high-quality instances serve as the foundation for multimodal sentiment analysis. Multimodal sentiment analysis is crucial in utilizing the complex context of acoustic and visual information to enrich text and improve the model's ability to predict sentiment. In this study, we aimed to enhance the contribution of nonverbal cues to sentiment analysis. From a modeling perspective, we devised the modality mixup strategy. By generating augmented samples with mixed acoustic and visual modalities, we enriched nonverbal behaviors and helped the model recognize the contributions of acoustic and visual behaviors. Our proposed approach using the MMix strategy is highly versatile and can adapt to various types of supervised and unsupervised data.

Improving text sentiment analysis with nonverbal modality remains a significant challenge today. Future studies can explore a simple, effective, and easily transferable end-to-end sentiment analysis model that leverages unsupervised data from the real world and can eventually be applied to emotion detection tasks in social media. Such a model can better utilize the rich audio and video information in social media, thereby enhancing the accuracy and efficiency of sentiment analysis. Additionally, this model can better handle nonverbal behaviors and extract valuable sentiment information from them, providing a more comprehensive perspective for text sentiment analysis.

5.2   Cross-Modal Sentiment Recognition Based on Hierarchical Grained and Acoustic Features

5.2.1   Introduction

When performing sentiment analysis, it can be easy to overlook the temporal fluctuations in acoustic data when simply extracting statistical features from the entire utterance. As a result, it is inadequate to solely extract acoustic features at the word-level and use convolution kernels of a CNN to learn acoustic local features.

In order to overcome these challenges, we introduced a hierarchical and feature-rich model. Drawing from the success of fine-grained representations in text-based emotion recognition, we sought to explore similar fine-grained representations in the acoustic domain. As a result, we modeled both frame-level and utterance-level structures of the acoustic data. Our model comprises three modules. Firstly, we utilize features extracted by LibROSA as input and employ a Gated Recurrent Unit network (F-GRU) to learn frame-level features containing temporal sequences, which are used to represent utterance-level features. We then use an utterance GRU (U-GRU) network to learn contextual information in dialogue and generate an utterance-level representation. Finally, we fuse this feature with statistical features extracted by openSMILE to obtain the final acoustic data representation feature. Our fusion phase is justifiable because the statistical feature extracted by openSMILE is based on utterance. We validated the effectiveness of our innovative work on the IEMOCAP and MELD datasets.

The remaining sections of this section are organized as follows. Part two presents the task definition. Part three provides a detailed overview of our hierarchical and feature-rich model. Part four outlines the experimental methodology and showcases the results. Finally, in part five, we summarize our conclusions and discuss future research directions.

5.2.2 Problem Definition

We are given a set of dialogues, D, where $D = [d_1, d_2, \ldots, d_L]$, with L representing the number of dialogues. Each dialogue, d_i, comprises a set of utterances, $u_{i,1}, u_{i,2}, \ldots, u_{i,N_i}$, where N_i is the number of utterances in each dialogue. Each utterance, $u_{i,j}$, contains a tuple, $\left\{ \left(A_j^O, A_j^L, E_j \right) \right\}_{j=1}^{N_i}$, where E_j represents the specific emotion for each utterance, such as anger, happiness, sadness, or neutrality. A_j^O represents a 1582-dimensional feature vector extracted by openSMILE, while $A_j^L \in R^{m \times n}$ represents the features extracted by LibROSA, where n is the time dimension, and m is the feature dimension.

Our goal is to predict emotions using various granular structures (frame-level and utterance-level) and features (A_j^O and A_j^L).

$$f_{\text{utt}}\left(A_j^L \right) \Rightarrow A_j^{\text{utt}} \tag{5.19}$$

$$f_{\text{emotion}}\left(A_j^{\text{utt}}, A_j^O \right) \Rightarrow E_j^{\text{pred}} \tag{5.20}$$

5.2.3 Methodology

In this section, we introduce the general framework of our hierarchical and feature-rich approach for acoustic emotion classification. Building upon the hierarchical GRU network [17], we have made further improvements. An overview of our general framework is depicted in Fig. 5.5, with the black dotted box representing feature extraction using LibROSA, and the green dotted box representing statistical feature extraction using openSMILE. To illustrate our approach, we use the Ses01F impro01 sample from the IEMOCAP dataset. Our main improvements include: (1) a frame-level module to extract acoustic frame-level features and learn contextual information within an utterance using a bidirectional GRU model; and (2) an utterance-level module to fuse the outputs of the frame-level structure with statistical features using a BiGRU network to learn the final utterance representation containing contextual information.

We extract frame-level acoustic features (A_j^L) using the Librosa [18] toolkit, which consist of 33 dimensions, including 20-dimensional Mel-frequency cepstral coefficients (MFCCs), 1-dimensional logarithmic fundamental frequency, and 12-dimensional constant-Q transform (CQT) features of the original acoustic data in time series. Additionally, we extract 1582 statistical acoustic features (A_j^O) using the openSMILE [5] toolkit.

In the frame-level phase, we employed a bidirectional GRU network [19] to extract feature vectors that contain frame-level information. For each A_j^L, we represent it as $A_j^L = [f_1, f_2, \ldots, f_{M_k}]$, where M_k is the number of frame windows (hop length) for each A_j^L. Using A_j^L as input, we learn frame-level embeddings in both forward and backward directions:

$$\overrightarrow{h_k} = \text{GRU}\left(A_j^L, \overrightarrow{h_{k-1}}\right) \tag{5.21}$$

$$\overleftarrow{h_k} = \text{GRU}\left(A_j^L, \overleftarrow{h_{k+1}}\right) \tag{5.22}$$

Building on prior work by Jiao [17], we calculate the self-attention of the hidden state in each direction. We then concatenate the frame-level embedding f_{emb}, h_k^r, $\overrightarrow{h_k}$, h_k^l, and $\overleftarrow{h_k}$, where $f_{\text{emb}} \in A_j^L$. The utterance-level embedding u_{emb} is obtained by performing max-pooling on the contextual frame embeddings. The tensor product is represented by the symbol $\otimes$.

$$h_k^r = \text{Softmax}\left(\overrightarrow{h_k} \otimes \overrightarrow{h_k}^T\right) \otimes \overrightarrow{h_k} \tag{5.23}$$

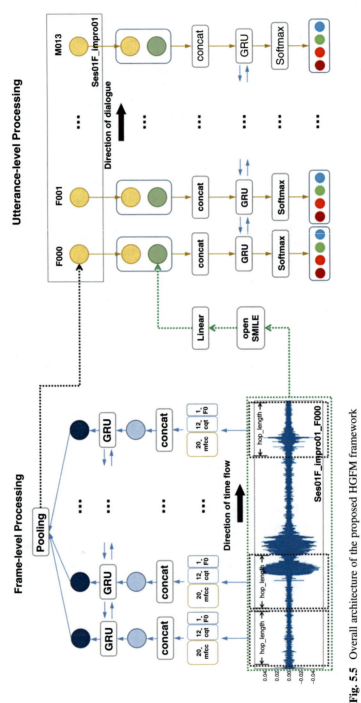

Fig. 5.5 Overall architecture of the proposed HGFM framework

$$h_k^l = \text{Softmax}\left(\overleftarrow{h_k} \otimes \overleftarrow{h_k}^{\text{T}}\right) \otimes \overleftarrow{h_k} \tag{5.24}$$

$$u_{\text{emb}} = \text{maxpool}\left(\text{concat}\left[f_{\text{emb}}, h_k^r, \overrightarrow{h_k}, h_k^l, \overleftarrow{h_k}\right]\right) \tag{5.25}$$

To control the statistical feature dimension, we employ a fully connected layer with the Tanh activation function. Here, $u_O \in A_j^O$. We set the number of hidden states in the GRU to be the same as the number of hidden layer neurons in linear transformation. Both u_{emb} and $u_{\text{op}} \in R^{1 \times d}$ have the same dimensions d for each.

$$u_{\text{op}} = \text{Tanh}(W_w \cdot u_O + b_w) \tag{5.26}$$

In the utterance-level phase, we concatenate u_{emb} and u_{op}, and then utilize a BiGRU network to learn contextual information in a dialogue.

$$A_j = \text{GRU}\left(\text{concat}\left[u_{\text{emb}}, u_{\text{op}}\right], \left(\overrightarrow{h_{k-1}}, \overleftarrow{h_{k+1}}\right)\right) \tag{5.27}$$

To obtain a probability distribution from the set of real vectors, we apply the Softmax activation function. We then utilize the Cross-Entropy loss as the objective function for optimization. We optimize the framework by minimizing the objective function as follows.

$$E_j^{\text{predA}} = \text{Softmax}\left(W_{\text{ER}} \cdot A_j + b_{\text{ER}}\right) \tag{5.28}$$

$$\text{loss} = -\sum_k y_k \log E_j^{\text{predA}} \tag{5.29}$$

5.2.4   Experiments

We conduct experiments on two datasets for dialogue emotion analysis. The number of utterances in each dataset is presented in detail in Table 5.6. As the MELD dataset has an imbalanced distribution of emotions, we assign weights to each emotion category during the experiment.

Table 5.6 Statistics of utterances for IEMOCAP and MELD datasets

Emotion							
Dataset	Happy/Joy	Anger	Sadness	Neutral	Surprise	Fear	Disgust
IEMCOAP	1636	1103	1084	1708	–	–	–
MELD	2308	1607	1002	6436	1636	358	361

Datasets

1. **IEMOCAP**: The IEMOCAP dataset [20] includes the following emotion labels: anger, happiness, sadness, neutral, excitement, frustration, fear, surprise, and others. To allow for a comparison with the state-of-the-art methods, we follow the approach taken by Zhou [21] and merge the happiness and excitement categories into a single "happy" category. Therefore, we focus on the four emotions of happy, angry, sadness, and neutral.
2. **MELD**: The MELD dataset [22] comprises approximately 13,000 utterances from 1,433 dialogues of the TV series Friends. We adopted the baseline experimental setup described in the section [22] for our experiments.

Compared Baselines

We assess and contrast various elements of our suggested HGFM methodology with several cutting-edge reference points:

1. **RNN** [23]: By utilizing a deep recurrent neural network (RNN) and a feature pooling strategy based on local attention.
2. **bcLSTM** [24]: A bidirectional contextual LSTM known as bcLSTM [24], leveraging multimodal features obtained through CNNs.
3. **MDNN** [21]: a generative multipath neural network that is semisupervised and employs acoustic features extracted through openSMILE.
4. **DialogueRNN** [25]: A neural network that is recurrent and monitors the states of the participants throughout the conversation.

Experimental Results

We measure the efficacy of our suggested hierarchical grained and feature model from two angles: (1) The accuracy performance, including weighted accuracy, unweighted accuracy [26], and F1-score. (2) The performance of each emotional category on the IEMOCAP dataset. The outcomes are presented in Tables 5.7 and 5.8. It is noteworthy that HGFM* solely utilizes the A_j^L feature for forecasting. The reported results are averaged over 10 experiments.

Table 5.7 The overall performance on IEMOCAP and MELD datasets comparison with the state-of-the-art

Method	IEMOCAP		MELD	
	WA	UWA	WA	UWA
RNN(2017 ICASSP)	63.5	58.8	38.4	20.6
bcLSTM(2017 ACL)	57.1	58.1	39.1	17.2
MDNN(2018 AAAI)	61.8	62.7	34.0	16.9
DialogueRNN(2019 AAAI)	65.8	66.1	41.8	22.7
HGFM*(our method)	62.6	68.2	41.4	19.9
HGFM(our method)	66.6	70.5	42.3	20.3

Table 5.8 Performance of each emotional category

Method	Angry	Happy	Sadness	Neutral
bcLSTM(2017 ACL)	58.37	60.45	61.35	52.31
DialogueRNN(2019 AAAI)	88.24	51.69	84.90	47.40
HGFM*(our method)	88.98	38.53	75.80	70.54
HGFM(our method)	88.84	54.37	72.51	67.36

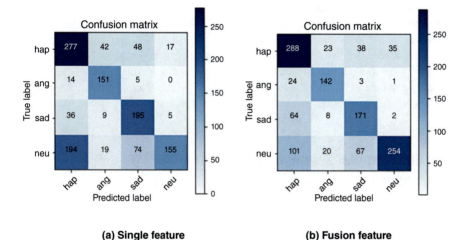

(a) Single feature   (b) Fusion feature

Fig. 5.6 The confusion matrix of different hierarchical features in emotional categories. (**a**) Single feature. (**b**) Fusion feature

We conducted a comparative study on four baselines using the IEOMCAP and MELD datasets. The results, highlighted in Table 5.7, were obtained by our experimentation. It is worth noting that IEMOCAP is a 4-way classification, while MELD is a 7-way classification. Our proposed HGFM model outperforms the state-of-the-art techniques in three evaluation metrics. The unweighted accuracy (UWA) showed a significant improvement of 4.4% on the IEMOCAP dataset. Based on the experimental findings, we analyzed the performance of our model from two perspectives: (1) The hierarchical grained design of our model allows it to more effectively learn the features of emotion recognition in acoustic data. This was validated by the enhanced accuracy in the experimental results. (2) As anticipated, the performance of the combined features was superior to that of the individual features, as shown in Table 5.8. The underlined results in the table were obtained by us. Although only neutral emotions achieved optimal performance in our model in terms of various emotions, the fusion of hierarchical features facilitated more balanced prediction accuracy for each emotion, which is a crucial factor in improving overall performance. This can be visualized in Fig. 5.6. We believe that acoustic data are more subjective, and people can use relatively peaceful sounds to express happiness. Our method is effective in capturing these implicit pieces of information.

5.2.5 Conclusion

The hierarchical grained and feature model was introduced in this section to address certain limitations of acoustic modality emotion recognition. Through a series of experiments, we have shown that this novel approach is highly effective in improving the accuracy of emotion recognition. The hierarchical granular structures enable us to capture more nuanced emotional cues that were previously overlooked, while the hierarchical feature helps us to obtain a more complete representation of the original acoustic data. By leveraging these two hierarchical structures, our proposed model can effectively classify emotions in a more accurate and comprehensive manner. Additionally, we believe that our model has the potential to be applied to other related research areas, such as speech recognition and natural language processing, where hierarchical structures may also play a crucial role in improving the accuracy of predictions.

5.3 Cross-Modal Sentiment Classification for Alignment Sequences

5.3.1 Introduction

In this section, we propose the Cross-Modal BERT (CM-BERT) approach to fine-tune the pretrained language representation model, Bidirectional Encoder Representations from Transformers (BERT), using multimodal information. While BERT has achieved state-of-the-art results on various natural language processing tasks, most previous works only fine-tuned it based on text data. Therefore, exploring how to improve BERT's representation by introducing multimodal information is still an important research direction. The proposed CM-BERT approach relies on cross-modal interaction between text and audio modality to fine-tune the pretrained BERT model. Masked multimodal attention, which dynamically adjusts the weight of words by combining text and audio modality information, is designed as the core unit of CM-BERT. We evaluated our method on two public multimodal sentiment analysis datasets, CMU-MOSI and CMU-MOSEI, and the experimental results demonstrate a significant improvement in performance on all metrics compared to previous baselines and text-only fine-tuning of BERT. Furthermore, we visualize the masked multimodal attention and prove that it can reasonably adjust the weight of words by incorporating audio modality information.

5.3.2   *Methodology*

Problem Definition

In this section, we propose a method for improving sentiment analysis performance by leveraging the interaction between text and audio modalities. Given a text sequence of word-piece tokens $T = [T_1, T_2, \ldots, T_n]$, where n is the sequence length, BERT's embedding layer appends a special classification embedding ([CLS]) before the input sequence. As a result, the output of the last encoder layer is a $n + 1$ length sequence, denoted as $X_t = [E_{[CLS]}, E_1, E_2, \ldots, E_n]$. To ensure consistency with the text modality, we append a zero vector before the word-level alignment audio features. The audio features are denoted as $X_a = [A_{[CLS]}, A_1, A_2, \ldots, A_n]$, where $A_{[CLS]}$ is a zero vector. Our approach aims to dynamically adjust the weight of each word by leveraging the interaction between X_t and X_a. By doing so, we can fine-tune the pretrained BERT model more effectively and improve sentiment analysis performance.

CM-BERT: Cross-Modal BERT

Figure 5.7 illustrates the architecture of our proposed CM-BERT model. The model takes in two inputs: the text sequence of word-piece tokens and the word-level alignment audio features. The text sequence is first passed through the pretrained BERT model, and the output of the last encoder layer is used as the text features, which is denoted as $X_t = [E_{[CLS]}, E_1, E_2, \ldots, E_n]$. The word-level alignment audio features, on the other hand, have a smaller dimension compared to X_t. To address this issue, we use a 1D temporal convolutional layer to adjust the dimension of the audio features to match that of the text features, following the approach used in [12]:

$$\left\{ \widehat{X_t}, \widehat{X_a} \right\} = \text{Conv1D} \left(\{ X_t, X_a \}, k_{\{t,a\}} \right) \tag{5.30}$$

The size of convolutional kernels for text and audio modalities is denoted as $k_{\{t, a\}}$. During training, the dimension of X_t is significantly higher than X_a. As a result, the dot products between X_t and X_a may grow large in magnitude, leading to extremely small gradients in the softmax function. To address this issue, we apply scaling to the text features $\widehat{X_t}$ to obtain $\widehat{X_t}'t$ and the audio features $\widehat{X_a}$ to obtain $\widehat{X_a}'$. This helps to prevent the dot products from growing too large and ensures stable training of the model:

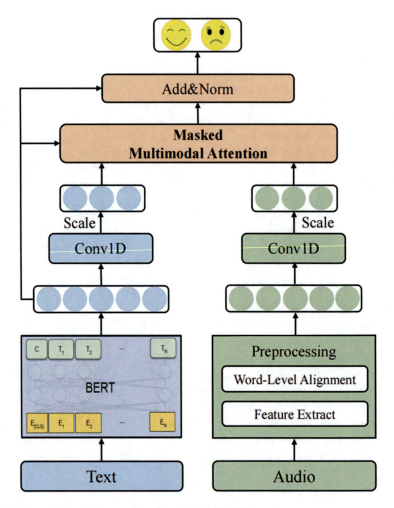

Fig. 5.7 Overview architecture of the Cross-Modal BERT Network

$$\widehat{X}_t' = \frac{\widehat{X}_t}{\sqrt{\left\|\widehat{X}_t\right\|_2}} \tag{5.31}$$

$$\widehat{X}_a' = \frac{\widehat{X}_a}{\sqrt{\left\|\widehat{X}_a\right\|_2}} \tag{5.32}$$

Once we have obtained X_t, $\widehat{X}_t'$, and $\widehat{X}_a'$, we use a masked multimodal attention mechanism to fully integrate the text and audio modalities and adjust the weight of words based on their performance in different modalities. The output of the masked multimodal attention is denoted as X_{Att}. To preserve the original structure of the data, we apply a residual connection to X_t and X_{Att}, following prior works [27, 28]. Next, the concatenated output is passed through a linear layer and a normalization layer. The final output of the last linear layer is $Y_l = [L_{\text{[CLS]}}, L_1, L_2, \ldots, L_n]$. The first token in the output sequence $L_{\text{[CLS]}}$ is a special classification token that is used to represent the entire input sequence. Since it is learned based on the information of all the other tokens in the sequence, it serves as a useful aggregate representation of the input data. We use this representation as input to a linear layer to generate the final prediction results.

Masked Multimodal Attention

The masked multimodal attention mechanism is a crucial aspect of the CM-BERT architecture, which allows us to leverage audio modality information to fine-tune the pretrained BERT model and adjust the weight of words. Figure 5.8 provides a visual representation of this mechanism. To begin with, we determine the weight of each word in both text and audio modalities. For the text modality, we define the Query Q_t and the Key K_t as $Q_t = K_t = \widehat{X}_t{}'$, where $\widehat{X}_t{}'$ represents the scaled text features. Similarly, for the audio modality, we define the Query Q_a and the Key K_a as

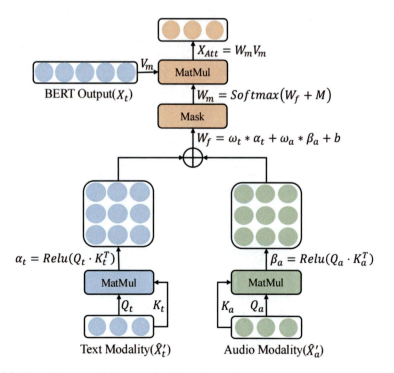

Fig. 5.8 The architecture of the masked multimodal attention

$Q_a = K_a = \widehat{X_a}'$, where $\widehat{X_a}'$ represents the scaled word-level alignment audio features. We then construct the text attention matrix a_t and the audio attention matrix β_a as:

$$a_t = \text{Relu}\left(Q_t K_t^{T}\right) \tag{5.33}$$

$$\beta_a = \text{Relu}\left(Q_a K_a^{T}\right) \tag{5.34}$$

Our goal is to dynamically adjust the weight of each word by leveraging the interaction between the text and audio modalities. To achieve this, we combine the text attention matrix a_t and the audio attention matrix β_a using a weighted sum. This results in the weighted fusion attention matrix W_f, which can effectively capture the multimodal information and provide more accurate sentiment analysis. The calculation of W_f involves a careful weighting of the text and audio modalities to ensure that both sources of information are given appropriate consideration.

$$W_f = w_t * a_t + w_a * \beta_a + b \tag{5.35}$$

We use weighting factors w_t and w_a to combine the text attention matrix a_t and audio attention matrix β_a, respectively, and add a bias term b to obtain the weighted fusion attention matrix W_f. To ensure that padding sequences do not affect the attention scores, we use a mask matrix M. The mask matrix represents the token positions as 0 and padding positions as $-\infty$ (after the softmax function, the attention score of padding positions becomes 0). Finally, we compute the multimodal attention matrix W_m as a combination of the weighted fusion attention matrix W_f and the mask matrix M. This helps to ensure that the model focuses only on the relevant parts of the input data and ignores the padding sequences:

$$W_m = \text{Softmax}\left(w_f + M\right) \tag{5.36}$$

Once we have computed the multimodal attention matrix, we use it to weigh the Value matrix V_m of the masked multimodal attention mechanism. The resulting product is the attention output matrix X_{Att}, which effectively captures the interaction between the text and audio modalities:

$$X_{\text{Att}} = W_m V_m \tag{5.37}$$

The Value matrix V_m is the output of the last encoder layer in the BERT model and is denoted as $V_m = X_t$.

5.3.3   Experiments

This section focuses on evaluating the performance of Cross-Modal BERT on two public multimodal sentiment analysis datasets: CMU-MOSI and CMU-MOSEI. Our experiments are presented in several stages. Firstly, we provide an overview of the datasets and experimental settings used in our study. Next, we describe the audio features and multimodal alignment techniques that we employ to enhance the model's performance. Finally, we introduce the evaluation metrics and baselines that we use to assess the effectiveness of our approach.

Datasets and Experimental Settings

We evaluate the effectiveness of our approach using two well-known multimodal sentiment analysis datasets: CMU Multimodal Opinion-level Sentiment Intensity (CMU-MOSI) [29] and CMU Multimodal Opinion Sentiment and Emotion Intensity (CMU-MOSEI) [11]. CMU-MOSI consists of 93 YouTube movie review videos, comprising 2199 utterances. Each utterance is annotated by 5 different workers and assigned a continuous sentiment score ranging from -3 (highly negative) to $+3$ (highly positive). To ensure the balance of positive and negative data and prevent speakers from appearing in both training and testing sets, we split the videos into 52, 10, and 31 for the training, validation, and test sets, respectively. This resulted in a total of 1284, 229, and 686 utterances across the three sets. Similarly, CMU-MOSEI is a multimodal sentiment and emotion analysis dataset, consisting of 23,454 movie review video clips from YouTube. In our experiments, we adopt a consistent strategy with previously published works [12, 30] to ensure comparability and reliability of the results.

Our proposed CM-BERT model utilizes the pretrained BERTBASE version, which consists of 12 transformer blocks. To prevent overfitting, we set the learning rate of the encoder layers to 0.01 and the learning rate of the other layers to 2e-5. We also freeze the parameters of the embedding layer to enhance model performance. During training, we use a batch size of 24 and set the maximum sequence length to 50. We train the model for 3 epochs using the Adam optimizer with mean-square error loss function. These hyperparameters are carefully selected to ensure optimal performance of the CM-BERT model on the sentiment analysis tasks.

Audio Features and Multimodal Alignment

In this study, we employ COVAREP [31] to extract audio features. Specifically, we represent each audio segment as a 74-dimensional feature vector, which includes a range of features such as 12 Mel-frequency cepstral coefficients (MFCCs), pitch and segmenting features, glottal source parameters, peak slope parameters, and maxima dispersion quotients. To obtain word-level alignment features, we leverage P2FA

[32] to determine the timestamps of each word in the audio segments. We then compute the average of the audio features within the corresponding word timestamps. To ensure that the audio features remain aligned with the text modality, we use zero vectors to pad the audio sequences.

Evaluation Metrics

In our experiments, we adopt the same evaluation metrics as in previous studies [30] to assess the performance of both the baselines and our proposed model. For the sentiment score classification task, we use 7-class accuracy (Acc_7), while for the binary sentiment classification task, we use both 2-class accuracy (Acc_2) and F_1 score (F_1). In the regression task, we evaluate the model performance using mean absolute error (MAE) and correlation (Corr) between the predicted and true labels. Higher values of Acc_7, Acc_2, F_1, and Corr indicate better model performance, while a lower MAE value is indicative of better performance. To ensure the robustness of our results, we randomly select five different seeds and report the average of the results obtained from 5 runs as the final evaluation metrics.

Baselines

We conduct a comparative analysis of the performance of CM-BERT against several previous models on multimodal sentiment analysis tasks. The models that we compare include

1. **EF-LSTM**: This model concatenates the multimodal inputs and employs a single LSTM to learn the contextual information. This model aims to integrate the audio and text information at an early stage and leverage the power of LSTM for sequence modeling.
2. **LMF** [9]: Another model that we compare against is Low-rank Multimodal Fusion (LMF), which is a method that employs low-rank weight tensors to efficiently fuse multimodal information without sacrificing performance. By using low-rank weight tensors, LMF is able to dramatically reduce the computational complexity of multimodal fusion, while still achieving significant improvements in performance. This allows for more efficient and effective modeling of the interaction between the text and audio modalities, and enhances the performance of the sentiment analysis task.
3. **MFN** [10]: Memory Fusion Network (MFN) is a model composed of three main components: the System of LSTMs, the Delta-memory Attention Network, and the Multiview Gated Memory. By explicitly accounting for interactions in the neural architecture and continuously modeling them over time, MFN is able to effectively fuse the multimodal information and enhance the performance of the sentiment analysis task.

4. **MARN** [33]: Multiattention Recurrent Network (MARN) is a model that leverages the Multiattention Block and the Long-short Term Hybrid Memory to effectively capture the interactions between different modalities. By using the Multiattention Block, MARN is able to selectively focus on the most relevant features within each modality, while the Long-short Term Hybrid Memory allows the model to capture the temporal dependencies between elements in the sequence.

5. **RMFN** [34]: Recurrent Multistage Fusion Network (RMFN) is a model that integrates the multistage fusion process with recurrent neural networks (RNNs) to effectively model both temporal and intramodal interactions. The multistage fusion process involves a series of fusion stages, where the information from different modalities is progressively fused to form a more comprehensive representation of the data.

6. **MFM** [35]: The Multimodal Factorization Model (MFM) is a model that is capable of factorizing multimodal representations into both multimodal discriminative factors and modality-specific generative factors. This approach enables each factor to focus on learning from a subset of the joint information across multimodal data and labels, which can enhance the performance of the sentiment analysis task.

7. **MCTN** [36]: The Multimodal Cyclic Translation Network (MCTN) is a model that is specifically designed to learn robust joint representations by translating between different modalities. This approach is particularly useful in situations where the available modalities may be limited or incomplete. During training, MCTN learns to translate between different modalities in a cyclic manner, allowing the model to effectively capture the interactions between the different modalities and learn a robust joint representation.

8. **MulT** [12]: The Multimodal Transformer (MulT) is a model that employs directional pairwise cross-modal attention to effectively capture the interactions between multimodal sequences across distinct time steps. Additionally, MulT is able to adapt the information from one modality to another in a latent manner.

9. **T-BERT** [4]: Bidirectional Encoder Representations from Transformers (BERT) is a language model that has achieved significant success in a wide range of natural language processing tasks, including sentiment analysis. BERT is trained on large amounts of text data, which allows it to learn representations that capture the complex relationships between words and phrases.

Results and Discussion

This section provides a detailed analysis of our experimental results and a comparison of our approach with previous work in the field of multimodal sentiment analysis. We also investigate the impact of introducing audio modality information and visualize the masked multimodal attention to gain insights into the inner workings of our model. Our experimental results demonstrate the effectiveness of our approach in capturing the complex interactions between different modalities and achieving state-of-the-art performance on multimodal sentiment analysis tasks. By comparing our approach with previous work, we are able to identify the strengths and weaknesses of different methods and gain insights into areas for future research.

We conducted an evaluation of the CM-BERT model on the CMU-MOSI dataset, and the results demonstrate that our model has achieved a new state-of-the-art performance on the task of multimodal sentiment analysis. The evaluation metrics for the binary sentiment classification task show that our model achieves an accuracy of 84.5% on Acc_2^h, which is a significant improvement compared to the baselines. Our model also achieves an improvement of 1.7–9.3% on F_1, which is another evaluation metric for the binary sentiment classification task. In the sentiment score classification task, the performance improvement of the CM-BERT model is even more significant. Our model achieves an accuracy of 44.9% on Acc_7^h, which is 4.9 to 12.1 percentage points higher than the baselines. In the regression task, our model reduces the mean absolute error (MAE) by about 0.142–0.294 and improves the correlation (Corr) by about 0.093–0.183. It is worth noting that the p-value for the student t-test between CM-BERT and T-BERT is far lower than 0.05 on all the metrics, indicating that our model significantly outperforms the baselines. Furthermore, it is noteworthy that all the baselines, except T-BERT, used information from text, audio, and video modalities, whereas our model only used text and audio modalities, yet it still achieved a new state-of-the-art result.

The experimental results demonstrate that the MulT model outperforms the other baseline models. This is mainly due to the fact that MulT extends the transformer architecture to the multimodal setting and employs attention mechanisms to adapt elements across different modalities. However, when compared to the T-BERT model, the latter achieves better performance by fine-tuning the pretrained BERT model to obtain better representations. In contrast, our proposed CM-BERT model extends the pretrained BERT model from unimodal to multimodal, and leverages audio modality information to effectively adjust the weight of words in the text modality. By utilizing this approach, the CM-BERT model is able to more comprehensively reflect the emotional state of the speaker and capture more nuanced sentiment characteristics through the interaction between text and audio modalities, leading to significant improvements across all evaluation metrics.

To further evaluate the effectiveness of our proposed method on multimodal language datasets beyond our training data, we conducted experiments on the CMU-MOSEI dataset. In order to facilitate comparison with previous work, we followed the evaluation metrics used in [37] and reported the Acc_2^h and F_1 scores for the top 3 models in Table 5.9. Our experiments revealed that while the MulT model achieves an Acc_2^h of 82.5% and an F_1 score of 82.3%, the T-BERT model performs better with an Acc_2^h of 83.0% and an F_1 score of 82.7%. However, our proposed CM-BERT model surpasses both MulT and T-BERT, achieving an Acc_2^h of 83.6% and an F_1 score of 83.6%.Compared with MulT and T-BERT, our model shows superior performance, with improvements of about 0.6–1.1% on Acc_2^h and 0.9–1.3% on F_1. These results demonstrate the generalization capability of our proposed

Table 5.9 Experimental results on CMU-MOSI dataset. The best results are highlighted in bold. h means higher is better and means lower is better. T text, A audio, V video

Model	Modality	Acc_7^h	Acc_2^h	F_1^h	MAE^l	$Corr^h$
EF-LSTM	T + A + V	33.7	75.3	75.2	1.023	0.608
LMF	T + A + V	32.8	76.4	75.7	0.912	0.668
MFN	T + A + V	34.1	77.4	77.3	0.965	0.632
MARN	T + A + V	34.7	77.1	77.0	0.968	0.625
RMFN	T + A + V	38.3	78.4	78.0	0.922	0.681
MFM	T + A + V	36.2	78.1	78.1	0.951	0.662
MCTN	T + A + V	35.6	79.3	79.1	0.909	0.676
MulT	T + A + V	40.0	83.0	82.8	0.871	0.698
T-BERT	T	41.5	83.2	83.2	0.784	0.774
CM-BERT	T + A	44.9	84.5	84.5	0.729	0.791

method and its ability to achieve superior performance on diverse multimodal language datasets.

Visualization of the Masked Multimodal Attention

In order to demonstrate the effectiveness of the masked multimodal attention mechanism, we conducted a visualization study to compare the text attention matrix A_t and the multimodal attention matrix W_m. By examining the differences in word weights, we can illustrate how the introduction of audio modality information enables the masked multimodal attention mechanism to adjust word weights more reasonably. To provide concrete examples, we selected three sentences from the MOSI dataset and visualized their text attention matrices and multimodal attention matrices in Fig. 5.9. Specifically, (a1), (a2), and (a3) represent the text attention matrices, while (b1), (b2), and (b3) represent the multimodal attention matrices. For instance, (a1) and (b1) correspond to the sentence "THERE ARE SOME FUNNY MOMENTS," (a2) and (b2) correspond to "I JUST WANNA SAY THAT I LOVE YOU," and (a3) and (b3) correspond to "I THOUGHT IT WAS FUN." In each of these three examples, we highlighted the most significant changes in word weights using red boxes. The color gradients in the visualizations indicate the relative levels of word importance.

We provided three examples to demonstrate the effectiveness of the masked multimodal attention mechanism in adjusting word weights and capturing important information through the interaction between text and audio modalities. In the first example, we examined the sentence "THERE ARE SOME FUNNY MOMENT" and compared its text attention matrix (a1) with its multimodal attention matrix (b1). We observed that the masked multimodal attention mechanism reduced the attention score of the word "ARE" and increased the attention on the words "SOME" and "MOMENTS." This adjustment is more reasonable and informative, as it captures the emotional content of the sentence more accurately. In the second example, we

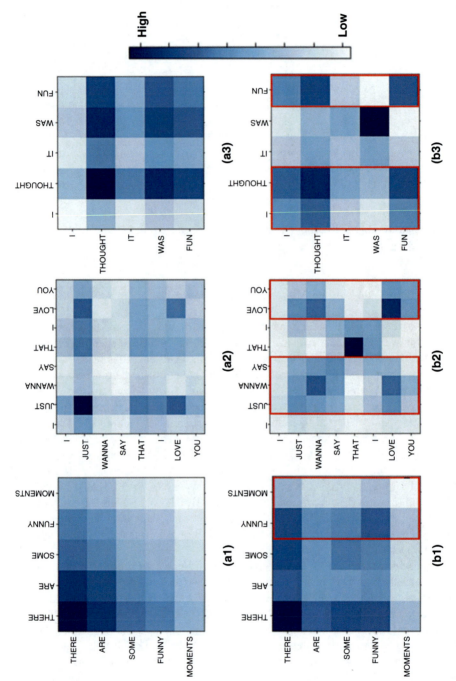

Fig. 5.9 Visualization of the attention matrices

analyzed the sentence "I JUST WANNA SAY THAT I LOVE YOU" and compared its text attention matrix (a2) with its multimodal attention matrix (b2). We found that the masked multimodal attention mechanism improved the weight between the words "LOVE" and "YOU" and reduced the weight between the words "JUST" and "THAT." These adjustments are consistent with human logic and help to capture richer emotional information while reducing the impact of noise. In the third example, we evaluated the sentence "I THOUGHT IT WAS FUN" and compared its text attention matrix (a3) with its multimodal attention matrix (b3). The masked multimodal attention mechanism again adjusted the weights of words in a more reasonable manner, with improvements observed between the word "I" and the words "THOUGHT" and "FUN." These adjustments are significant, as they capture the emotional tone of the speaker and help to improve the accuracy of sentiment analysis.

5.3.4 Conclusion

In this section, we introduce a new multimodal sentiment analysis model named Cross-Modal BERT (CM-BERT). Unlike previous works, our model extends the pretrained BERT model from unimodal to multimodal by incorporating audio modality information to fine-tune BERT and obtain better representations. The core unit of CM-BERT is the masked multimodal attention mechanism, which dynamically adjusts the weight of words through intermodality interaction between text and audio modalities. Our experimental results demonstrate that CM-BERT significantly outperforms previous baselines and text-only fine-tuning of BERT on the CMU-MOSI and CMU-MOSEI datasets. Additionally, we visualize the attention matrices to show how the masked multimodal attention mechanism can adjust the word weights more reasonably after introducing the audio modality. Importantly, CM-BERT is not limited to two modalities and can be extended to text and video modalities, as well as more than two modalities. In the future, we plan to explore methods for aligning different modalities, as most of the multimodal data in the real world are usually unaligned. This will help to further improve the performance of multimodal sentiment analysis and enable the effective analysis of diverse multimodal data in a wide range of applications.

5.4 Summary

This chapter explores multimodal sentiment analysis, emphasizing the importance of incorporating audio and visual modalities in social media to enrich text-based sentiment analysis. This chapter proposes approaches such as modality mixup to enhance the recognition of nonverbal cues and improve the model's performance. It also introduces a hierarchical grained and feature model for addressing limitations in acoustic modality emotion recognition. Additionally, this chapter presents the Cross-Modal BERT (CM-BERT) model, which extends the pretrained BERT model to incorporate audio modality information. Overall, this chapter highlights the significance of multimodal sentiment analysis, introduces innovative approaches, and suggests future research directions for improving sentiment analysis in diverse multimodal data.

References

1. Hochreiter S, Schmidhuber J (1997) Long short-term memory. Neural Comput 9(8):1735–1780
2. Verma V, Kawaguchi K, Lamb A et al (2022) Interpolation consistency training for semi-supervised learning. Neural Netw 145:90–106
3. Yu W, Xu H, Meng F, et al (2020) Ch-sims: A chinese multimodal sentiment analysis dataset with fine-grained annotation of modality. Proceedings of the 58th Annual Meeting of the Association for Computational Linguistics, 3718–3727
4. Devlin J, Chang MW, Lee K, et al (2019) Bert: Pre-training of deep bidirectional transformers for language understanding. Proceedings of the 2019 Conference of the North American Chapter of the Association for Computational Linguistics: Human Language Technologies, (1):4171–4186
5. Eyben F, Wöllmer M, Schuller B (2010) Opensmile: the munich versatile and fast open-source audio feature extractor. Proceedings of the 18th Association for Computing Machinery international conference on Multimedia, 1459–1462
6. Tao R, Pan Z, Das RK, et al (2021) Is someone speaking? Exploring long-term temporal features for audio-visual active speaker detection. Proceedings of the 29th Association for Computing Machinery International Conference on Multimedia, 3927–3935
7. Jennifer Williams, Steven Kleinegesse, Ramona Comanescu, et al (2018) Recognizing emotions in video using multimodal DNN feature fusion. Proceedings of Grand Challenge and Workshop on Human Multimodal Language, 11–19
8. Zadeh A, Chen M, Poria S, et al (2017) Tensor fusion network for multimodal sentiment analysis. Proceedings of the 2017 Association for Computational Linguistics Conference on Empirical Methods in Natural Language Processing, 1103–1114
9. Liu Z, Shen Y, Lakshminarasimhan VB, et al (2018) Efficient low-rank multimodal fusion with modality-specific factors. Proceedings of the 56th Annual Meeting of the Association for Computational Linguistics, (1):2247–2256
10. Zadeh A, Liang PP, Mazumder N, et al (2018) Memory fusion network for multi-view sequential learning. Proceedings of the 32nd Association for the Advancement of Artificial Intelligence Conference on Artificial Intelligence, 5634–5641
11. Zadeh AAB, Liang PP, Poria S, et al (2018) Multimodal language analysis in the wild: Cmu-mosei dataset and interpretable dynamic fusion graph. Proceedings of the 56th Annual Meeting of the Association for Computational Linguistics, (1): 2236–2246
12. Tsai YHH, Bai S, Liang PP, et al (2019) Multimodal transformer for unaligned multimodal language sequences. Proceedings of the 57th Annual Meeting of the Association for Computational Linguistics, 6558–6569
13. Rahman W, Hasan MK, Lee S, et al (2020) Integrating multimodal information in large pretrained transformers. Proceedings of the 58th Annual Meeting of the Association for Computational Linguistics, 2359–2369
14. Hazarika D, Zimmermann R, and Poria S (2020) Misa: Modality-invariant and-specific representations for multimodal sentiment analysis[C]. Proceedings of the 28th Association for Computing Machinery International Conference on Multimedia, 1122–1131

15. Han W, Chen H, Poria S (2021) Improving multimodal fusion with hierarchical mutual information maximization for multimodal sentiment analysis. Proceedings of the 2021 Conference on Empirical Methods in Natural Language Processing, 9180–9192

16. Wenmeng Y, Hua X, Yuan Z et al (2021) Learning modality-specific representations with self-supervised multi-task learning for multimodal sentiment analysis. Proc Assoc Advanc Artific Intellig Conf Artific Intellig 35(12):10790–10797

17. Jiao W, Yang H, King I, et al (2019) HIGRU: Hierarchical gated recurrent units for utterance-level emotion recognition. Proceedings of the 2019 Conference of the North American Chapter of the Association for Computational Linguistics: Human Language Technologies, 1: 397–406

18. McFee B, Raffel C, Liang D, et al (2015) Librosa: Audio and music signal analysis in python. Proceedings of the 14th python in science conference, 18–25

19. Cho K, van Merriënboer B, Gulcehre C, et al (2014) Learning phrase representations using rnn encoder-decoder for statistical machine translation. Proceedings of the 2014 Conference on Empirical Methods in Natural Language Processing, 1724-1734

20. Busso C, Bulut M, Lee CC et al (2008) IEMOCAP: interactive emotional dyadic motion capture database. Lang Resour Eval 42(4):335–359

21. Zhou S, Jia J, Wang Q, et al (2018) Inferring emotion from conversational voice data: A semi-supervised multi-path generative neural network approach. Proceedings of the 32nd Association for the Advancement of Artificial Intelligence Conference on Artificial Intelligence and 20th Innovative Applications of Artificial Intelligence Conference and 8th Association for the Advancement of Artificial Intelligence Symposium on Educational Advances in Artificial Intelligence, (72): 579–586

22. Poria S, Hazarika D, Majumder N, et al (2019) MELD: A Multimodal Multi-Party Dataset for Emotion Recognition in Conversation. Proceedings of the 57th Annual Meeting of the Association for Computational Linguistics, 527–536

23. Mirsamadi S, Barsoum E, Zhang C (2017) Automatic speech emotion recognition using recurrent neural networks with local attention[C]. Proceedings of the 2017 Institute of Electrical and Electronics Engineers International Conference on Acoustics, Speech and Signal Processing, 2227–2231

24. Poria S, Cambria E, Hazarika D, et al (2017) Context-dependent sentiment analysis inuser-generated videos. Proceedings of the 55th Annual Meeting of the Association for Computational Linguistics, 1: 873–883

25. Majumder N, Poria S, Hazarika D et al (2019) Dialoguernn: an attentive rnn for emotion detection in conversations. Proc Assoc Advanc Artific Intellig Confer Artific Intellig 33(837): 6818–6825

26. Rozgić V, Ananthakrishnan S, Saleem S, et al (2012) Ensemble of SVM trees for multimodal emotion recognition. Proceedings of the 2012 Asia Pacific Signal and Information Processing Association Annual Summit and Conference, 1–4

27. Gao S, Chen X, Li P, et al (2019) How to write summaries with patterns? learning towards abstractive summarization through prototype editing. Proceedings of the 2019 Conference on Empirical Methods in Natural Language Processing and the 9th International Joint Conference on Natural Language Processing, 3741–3751

28. Vaswani A, Shazeer N, Parmar N, et al (2017) Attention is all you need. Proceedings of the 31st International Conference on Neural Information Processing Systems, 5998–6008

29. Zadeh A, Zellers R, Pincus E (2016) Multimodal sentiment intensity analysis in videos: facial gestures and verbal messages[J]. IEEE Intell Syst 31(6):82–88

30. Wang Y, Shen Y, Liu Z et al (2019) Words can shift: dynamically adjusting word representations using nonverbal behaviors. Proc Assoc Advanc Artific Intellig Conf Artific Intellig 33(1): 7216–7223

31. Degottex G, Kane J, Drugman T, et al (2014) COVAREP—A collaborative voice analysis repository for speech technologies. Proceedings of the 2014 Institute of Electrical and Electronics Engineers international conference on acoustics, speech and signal processing, 960–964

32. Yuan J, Liberman M (2008) Speaker identification on the SCOTUS corpus. J Acoust Soc Am 123(5):3878
33. Dalal N, Triggs B (2005) Histograms of oriented gradients for human detection. Proceedings of the 2005 Institute of Electrical and Electronics Engineering Computer Society Conference on Computer Vision and Pattern Recognition, 1: 886–893
34. Liang PP, Liu Z, Zadeh AAB, et al (2018) Multimodal language analysis with recurrent multistage fusion. Proceedings of the 2018 Conference on Empirical Methods in Natural Language Processing, 150–161
35. Tsai YHH, Liang PP, Zadeh A, et al (2019) Learning factorized multimodal representations. Proceedings of the 2019 International Conference on Representation Learning.
36. Pham H, Liang PP, Manzini T et al (2019) Found in translation: learning robust joint representations by cyclic translations between modalities. Proc Assoc Advanc Artific Intellig Conf Artific Intellig 33:6892–6899
37. George T, Fabien R, Raymond B, et al (2016) Adieu features? end-to-end speech emotion recognition using a deep convolutional recurrent network. Proceedings of the 2016 Institute of Electrical and Electronics Engineers international conference on acoustics, speech and signal processing, 5200–5204

Chapter 6
Multimodal Sentiment Analysis

Abstract This chapter discusses the increasing importance of Multimodal Sentiment Analysis (MSA) in social media data analysis. It introduces the challenge of Representation Learning and proposes a self-supervised label generation module and joint training approach to improve multimodal representations. This chapter presents experimental results demonstrating the reliability of autogenerated unimodal supervisions, offering a promising solution for enhancing MSA without manual annotations. The second section addresses the challenge of missing data in Multimodal Sentiment Analysis (MSA). It presents a transformer-based feature reconstruction network (TFR-Net) that utilizes attention-based extractors and a reconstruction module to handle random missing modality features in nonaligned sequences. Experimental evaluations showcase the model's robustness and high performance across various missing modality combinations and degrees of missingness.

6.1 Multimodal Sentiment Analysis Model Based on Self-Supervised Multitask Learning

6.1.1 Introduction

In recent years, Multimodal Sentiment Analysis (MSA) has gained increasing attention [1–3] due to its superior performance compared to unimodal sentiment analysis when dealing with social media data. As the amount of user-generated online content continues to grow, MSA has been applied to various domains such as risk management, video understanding, and video transcription.

Representation Learning is a significant and challenging task in multimodal learning. Effective modality representations should capture both the consistency and difference between modalities. However, existing methods that rely on unified multimodal annotation have limited capacity to capture differentiated information. Moreover, obtaining additional unimodal annotations can be time-consuming and labor-intensive.

To address these challenges, we propose a label generation module based on self-supervised learning to acquire independent unimodal supervisions. We then jointly train the multimodal and unimodal tasks to learn consistency and difference, respectively. During the training stage, we introduce a weight-adjustment strategy to balance the learning progress among different subtasks. This strategy guides the subtasks to focus on samples with larger differences between modality supervisions.

We conduct extensive experiments on three public multimodal baseline datasets and demonstrate the reliability and stability of our autogenerated unimodal supervisions. Our approach shows promise in improving the quality of multimodal representations without the need for additional manual annotations.

6.1.2   Methodology

In this section, we will provide a thorough explanation of the Self-Supervised Multitask Multimodal sentiment analysis network, or Self-MM for short. The primary aim of the Self-MM is to acquire highly informative unimodal representations by simultaneously learning one multimodal task and three unimodal subtasks. Unlike the multimodal task, which uses human-annotated labels, the unimodal subtasks employ a self-supervised approach to autogenerate their labels. To simplify references in subsequent sections, we will use "m-labels" to denote the human-annotated multimodal labels and "u-labels" to indicate the autogenerated unimodal labels.

Task Setup

The aim of Multimodal Sentiment Analysis (MSA) is to determine sentiment using a combination of different types of signals, such as text (I_t), audio (I_a), and visual (I_v) inputs. Generally, MSA can be considered either a regression or a classification task. In this particular work, we approach it as a regression task. The Self-Supervised Multitask Multimodal sentiment analysis network (Self-MM) takes in I_t, I_a, and I_v as inputs and produces a single sentiment intensity result, $\hat{y}_m \in R$. To aid in representation learning during the training stage, Self-MM also generates three additional unimodal outputs, $\hat{y}_s \in R$, where $s \in \{t, a, v\}$. Despite having multiple outputs, we only use $\hat{y}_m$ as the final predictive result.

As depicted in Fig. 6.1, the Self-Supervised Multitask Multimodal sentiment analysis network (Self-MM) comprises both a multimodal task and three separate unimodal subtasks. The $\hat{y}_m$, $\hat{y}_t$, $\hat{y}_a$, and $\hat{y}_v$ refer to the predictive outputs of the multimodal and unimodal tasks, respectively. The y_m is the human-annotated multimodal label, whereas the y_t, y_a, and y_v represent the autogenerated unimodal supervision from the self-supervised strategy. Ultimately, $\hat{y}_m$ is utilized as the sentiment output. To

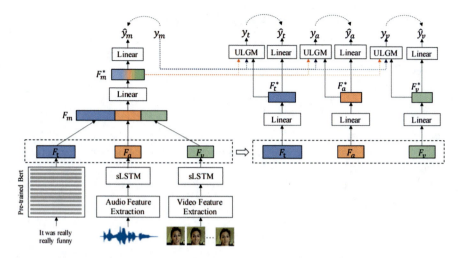

Fig. 6.1 The overall architecture of Self-MM

facilitate representation learning, we employ a hard-sharing strategy between the bottom representation learning network of the multimodal and different unimodal tasks.

Multimodal Task

To handle the multimodal task, we utilize a classical architecture for multimodal sentiment analysis. This architecture comprises three primary components: the feature representation module, the feature fusion module, and the output module. In the text modality, we capitalize on the exceptional performance of pretrained language models and implement the pretrained 12-layers BERT to extract sentence representations. We empirically determine that the first-word vector in the last layer provides the optimal representation for the entire sentence, denoted as F_t.

$$F_t = \text{BERT}\left(I_t; \theta_t^{\text{bert}}\right) \in R^{d_t} \qquad (6.1)$$

To handle the audio and vision modalities, we first utilize pretrained ToolKits to extract the initial vector features, $I_a \in R^{l_a \times d_a}$ and $I_v \in R^{l_v \times d_v}$, from the raw data. Here, l_a and l_v represent the respective sequence lengths for the audio and vision inputs. We then employ a single directional Long Short-Term Memory (sLSTM) [4] to capture the timing characteristics of the sequences. Finally, we extract the end-state hidden vectors as the complete sequence representations.

$$F_a = \text{sLSTM}\left(I_a; \theta_a^{\text{lstm}}\right) \in R^{d_a} \qquad (6.2)$$

$$F_v = \text{sLSTM}\left(I_v; \theta_v^{\text{lstm}}\right) \in R^{d_v} \tag{6.3}$$

Next, we merge all the unimodal representations and project them into a lower-dimensional space of R^{d_m} through concatenation:

$$F_m^* = \text{Re}\,LU\left(W_{l1}^{mT}[F_t; F_a; F_v] + b_{l1}^m\right) \tag{6.4}$$

where we leverage the ReLU activation function and the matrix $W_{l1}^m \in R^{(d_t + d_a + d_v) \times d_m}$ to project the merged unimodal representations into a lower-dimensional space of R^{d_m}.

Finally, we utilize the fused representation F_m^* to make predictions for the multimodal sentiment analysis task.

$$\hat{y}_m = W_{l2}^{mT} F_m^* + b_{l2}^m \tag{6.5}$$

where $W_{l2}^m \in R^{d_m \times 1}$.

Unimodal Task

The unimodal tasks utilize the same modality representations as the multimodal task. To address the issue of varying dimensions across different modalities, we perform a projection of the features into a novel feature space. Subsequently, linear regression is employed to obtain the unimodal outcomes.

$$F_s^* = \text{Re}\,LU\left(W_{l1}^{sT} F_s + b_{l1}^s\right) \tag{6.6}$$

$$\hat{y}_s = W_{l2}^{sT} F_s^* + b_{l2}^s \tag{6.7}$$

where $s \in \{t, a, v\}$.

To facilitate the training of the unimodal task, a module known as the Unimodal Label Generation Module (ULGM) has been developed. This module is specifically designed to generate labels.

$$y_s = \text{ULGM}\left(y_m, F_m^*, F_s^*\right) \tag{6.8}$$

where $s \in \{t, a, v\}$.

Ultimately, we engage in joint learning of the multimodal task and the three unimodal tasks, while being supervised by both m-labels and u-labels. Notably, these unimodal tasks are solely present during the training phase. Consequently, we utilize $\hat{y}_m$ as the ultimate output result.

ULGM

The Unimodal Label Generation Module (ULGM) is responsible for producing unimodal supervision values by relying on both multimodal annotations and modality representations. To prevent any unwanted interference with the network parameter updates, the ULGM is designed as a nonparameter module. In general, unimodal supervision values are closely linked to multimodal labels. As such, the ULGM calculates the offset based on the relative distance between the modality representations and the class centers, as depicted in Fig. 6.2. The multimodal representation F_m^* is in closer proximity to the positive center (m-pos), while the unimodal representation is closer to the negative center (s-neg). Therefore, the unimodal supervision value y_s is adjusted with a negative offset δ_{sm}, which is added to the multimodal label y_m.

Relative Distance Value

Given that distinct modality representations exist in separate feature spaces, utilizing the absolute distance value may not yield precise results. As a remedy, we introduce the relative distance value, which is independent of space differences. To start, during the training phase, we establish and retain the positive center (C_i^p) and the negative center (C_i^n) for various modality representations:

$$C_i^p = \frac{\sum_{j=1}^{N} I(y_i(j) > 0) \cdot F_{ij}^g}{\sum_{j=1}^{N} I(y_i(j) > 0)} \tag{6.9}$$

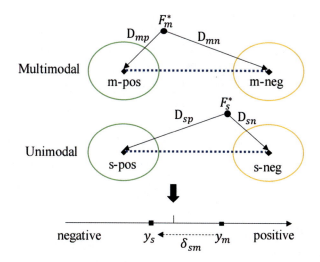

Fig. 6.2 Unimodal label generation example

$$C_i^n = \frac{\sum\limits_{j=1}^{N} I(y_i(j) < 0) \cdot F_{ij}^g}{\sum\limits_{j=1}^{N} I(y_i(j) < 0)} \tag{6.10}$$

where i belongs to $\{m, t, a, v\}$, N denotes the total number of training samples, and $I(\cdot)$ represents an indicator function. F_{ij}^g refers to the global representation of the jth sample in the ith modality.

Regarding modality representations, we employ L2 normalization to determine the distance between F_i^* and class centers:

$$D_i^p = \frac{\| F_i^* - C_i^p \|_2^2}{\sqrt{d_i}} \tag{6.11}$$

$$D_i^n = \frac{\| F_i^* - C_i^n \|_2^2}{\sqrt{d_i}} \tag{6.12}$$

where i belongs to $\{m, t, a, v\}$. Furthermore, d_i refers to the representation dimension, while the scale factor represents a constant value.

Subsequently, we establish the relative distance value, which assesses the distance between the modality representation and both the positive and negative centers, in relation to each other.

$$\alpha_i = \frac{D_i^n - D_i^p}{D_i^p + \epsilon} \tag{6.13}$$

where $i \in \{m, t, a, v\}$. ϵ denotes a small numerical value.

Shifting Value

It is reasonable to assume that the final results are positively influenced by the value of α_i. In order to establish a connection between the supervisions and the predicted values, we examine the following two relationships.

$$\frac{y_s}{y_m} \propto \frac{\hat{y}_s}{\hat{y}_m} \propto \frac{\alpha_s}{\alpha_m} \Rightarrow y_s = \frac{\alpha_s * y_m}{\alpha_m} \tag{6.14}$$

$$y_s - y_m \propto \hat{y}_s - \hat{y}_m \propto \alpha_s - \alpha_m \Rightarrow y_s = y_m + \alpha_s - \alpha_m \tag{6.15}$$

where $s \in \{t, a, v\}$.

To circumvent the "zero value problem," we introduce Eq. (6.15). In Eq. (6.14), if y_m is equal to zero, the generated unimodal supervision values y_s will always be zero. By taking into account the aforementioned relationships, we can obtain unimodal supervisions through an equal-weight summation:

$$
\begin{aligned}
y_s &= \frac{y_m * \alpha_s}{2\alpha_m} + \frac{y_m + \alpha_s - \alpha_m}{2} \\
&= y_m + \frac{\alpha_s - \alpha_m}{2} * \frac{y_m + \alpha_m}{\alpha_m} \\
&= y_m + \delta_{sm}
\end{aligned}
\tag{6.16}
$$

where s belongs to $\{t, a, v\}$. The offset value of the unimodal supervisions in relation to the multimodal annotations is denoted as δ_{sm}, which is calculated using the following formula: $\delta_{sm} = \frac{\alpha_t - \alpha_m}{2} * \frac{y_m + \alpha_m}{\alpha_m}$.

Momentum-based Update Policy

As a result of the varying nature of modality representations, the u-labels that are generated using Eq. (6.16) may not be entirely stable. To address this issue, we implement a momentum-based update policy, which merges the newly generated value with the historical values to alleviate any negative effects.

$$
y_s^{(i)} = \begin{cases}
y_m & i = 1 \\
\frac{i-1}{i+1} y_s^{(i-1)} + \frac{2}{i+1} y_s^i & i > 1
\end{cases}
\tag{6.17}
$$

In this case, s belongs to $\{t, a, v\}$. The variable y_s^i refers to the newly generated u-labels at the ith epoch, while $y_s^{(i)}$ represents the final u-labels after the ith epoch.

Algorithm 6.1: Unimodal Supervisions Update Policy

Input: unimodal inputs I_t, I_a, I_v, m-labels y_m
Output: u-labels $y_t^{(i)}, y_a^{(i)}, y_v^{(i)}$ where i means the number of training epochs
1: Initialize model parameters $M(\theta; x)$
2: Initialize u-labels $y_t^{(1)} = y_m, y_a^{(1)} = y_m, y_v^{(1)} = y_m$
3: Initialize global representations $F_t^g = 0, F_a^g = 0, F_v^g = 0, F_m^g = 0$
4: **for** $n \in [1, end]$ **do**
5:   **for** minibatch in data loader **do**
6:     Compute minibatch modality representations $F_t^*, F_a^*, F_v^*, F_m^*$
7:     Compute loss L using Eq. (6.18)
8:     Compute parameters gradient $\frac{\partial L}{\partial \theta}$
9:     Update model parameters: $\theta = \theta - \eta \frac{\partial L}{\partial \theta}$
10:      **if** $n \neq 1$ **then**
11:        Compute relative distance values $\alpha_m, \alpha_t, \alpha_a$, and α_v using Eqs.(6.9)–(6.13)
12:        Compute y_t, y_a, y_v using Eq.(6.16)
13:        Update $y_t^{(n)}, y_a^{(n)}, y_t^{(n)}$ using Eq.(6.17)

(continued)

14:		end if
15:		Update global representations F_s^g using F_s^*, where $s \in \{m, t, a, v\}$
16:	**end if**	
17:	**end if**	

To be more precise, assuming that the total number of epochs is n, we can determine that the weight of y_s^i is $\frac{2i}{(n)(n+1)}$. This implies that the weight of the u-labels that are generated at later epochs is greater than that of the earlier ones. This aligns with our intuition, as the unimodal labels that are generated are the summation of all the previous epochs, and they become more stable after a sufficient number of iterations (approximately 20 in our experiments). As a result, the training process of the unimodal tasks gradually stabilizes. The unimodal labels update policy is depicted in Algorithm 6.1.

Optimization Objectives

Finally, we utilize the L1 Loss as the fundamental optimization objective. For the unimodal tasks, the weight of the loss function is determined by the difference between the u-labels and the m-labels. This implies that the network should emphasize the samples with a greater difference between the two labels.

$$L = \frac{1}{N} \sum_i^N \left(\left| \hat{y}i - y_m^i \right|_m + \sum_s^{\{t, a, v\}} W_s^i * \left| \hat{y}i - y_s^{(i)} \right|_s \right) \tag{6.18}$$

where N is the number of training samples. $W_s^i = \tanh\left(\left| y_s^{(i)} - y_m \right| \right)$ is the weight of ith sample for auxiliary task s.

6.1.3 Experiments

This section outlines our experimental setup, which includes the datasets used, the baseline models, and the evaluation metrics.

Datasets

Our study employs three publicly available datasets for multimodal sentiment analysis, namely MOSI [4], MOSEI [5], and SIMS [6]. The key statistics for these datasets are presented in Table 6.1. Below, we provide a brief overview of each dataset.

1. **MOSI:** One of the most widely used benchmark datasets for multimodal sentiment analysis (MSA) is the CMU-MOSI dataset. It contains 2199 brief monologue video clips that are extracted from 93 movie review videos on YouTube.
   Each sample is assigned a sentiment score by human annotators, ranging from -3
   (strongly negative) to 3 (strongly positive).
2. **MOSEI:** Compared to CMU-MOSI, the CMU-MOSEI dataset features a larger
   number of utterances, a greater diversity of samples, speakers, and topics. It
   comprises 23,453 annotated video segments (utterances), which are sourced
   from 5000 videos, encompassing 1000 distinct speakers and 250 different topics.
3. **SIMS:** The SIMS dataset is a unique benchmark for multimodal sentiment
   analysis (MSA) in the Chinese language, featuring finely-grained annotations of
   modality. The dataset comprises 2281 high-quality video clips that are meticulously selected from a range of movies, TV series, and variety shows, showcasing
   spontaneous expressions, diverse head poses, occlusions, and illuminations. Each
   sample is assigned a sentiment score by human annotators, ranging from -1
   (strongly negative) to 1 (strongly positive).

Baselines

Our study aims to thoroughly assess the effectiveness of Self-MM, and to this end,
we conduct a comprehensive comparison with a range of baseline models and state-
of-the-art methods that have been developed for multimodal sentiment analysis.

1. **TFN:** The Tensor Fusion Network (TFN) leverages outer-product operations to
   compute a multidimensional tensor that effectively captures the interactions
   between unimodal, bimodal, and trimodal features.
2. **LMF:** Low-rank Multimodal Fusion (LMF) [7] is an enhanced version of TFN
   that utilizes a low-rank fusion technique for multimodal tensors in order to
   improve computational efficiency without compromising performance.
3. **MFN:** The Memory Fusion Network (MFN) [8] is specifically designed to model
   both view-specific and cross-view interactions in a continuous manner, and to
   summarize them over time using a Multiview Gated Memory mechanism.
4. **MFM:** The Multimodal Factorization Model (MFM) [9] is a method that learns
   generative representations to capture modality-specific generative features, as
   well as discriminative representations that facilitate classification.
5. **RAVEN:** The Recurrent Attended Variation Embedding Network (RAVEN)
   [10] is an attention-based model that incorporates nonverbal signals to adjust
   word embeddings. This is achieved through a process of re-adjustment, which
   enables the model to better capture the variations in the input data.

Table 6.1 Dataset statistics in MOSI, MOSEI, and SIMS

Dataset	# Train	# Valid	# Test	# All
MOSI	1284	229	686	2199
MOSEI	16,326	1871	4659	22,856
SIMS	1368	456	457	2281

6. **MulT**: The Multimodal Transformer (MulT) represents a significant advancement in the multimodal transformer architecture, as it incorporates directional pairwise crossmodal attention. This innovative approach facilitates the seamless translation of information between different modalities through the use of directional pairwise cross-attention.
7. **MAG-BERT:** The Multimodal Adaptation Gate for Bert (MAG-BERT) [11] represents a significant improvement over RAVEN, particularly when working with aligned data. This is achieved by applying a multimodal adaptation gate at various layers of the BERT backbone.
8. **MISA:** The Modality-Invariant and Modality-Specific Representations (MISA) [12] is a novel technique that leverages a combination of different loss functions, including distributional similarity, orthogonal loss, reconstruction loss, and task prediction loss, to learn both modality-invariant and modality-specific representations. This enables the model to effectively capture the nuances of different modalities while also identifying commonalities across them.

Basic Settings

1. **Experimental Details**: In our experiments, we adopt the Adam optimizer and set the initial learning rate to 5e-5 for Bert and 1e-3 for other parameters. To ensure a fair comparison, we train our Self-MM model and compare it to two state-of-the-art approaches, MISA and MAG-BERT, by conducting five independent runs for each model and presenting the average performance.
2. **Evaluation Metrics**: As per previous studies [11, 12], we present our experimental findings in two distinct formats: classification and regression. In the classification section, we report both Weighted F1 score (F1-Score) and binary classification accuracy (Acc-2). Specifically, for the MOSI and MOSEI datasets, we assess Acc-2 and F1-Score using two separate methods: negative/nonnegative (nonexcluding zero) [1] and negative/positive (excluding zero) [2]. In the regression section, we present the Mean Absolute Error (MAE) and Pearson correlation (Corr) as our performance metrics. It is important to note that, with the exception of MAE, higher values indicate superior performance for all metrics.

Results and Analysis

This section provides a comprehensive analysis and discussion of our experimental results.

Table 6.2 presents a comparative analysis of our experimental results on the MOSI and MOSEI datasets. Models denoted with (B) indicate the use of BERT-based language features, while those marked with an asterisk were reproduced under the same conditions. In the Acc-2 and F1-Score metrics, the left of the slash "/" represents the "negative/nonnegative" calculation, while the right represents the "negative/positive" calculation. To ensure a fair comparison, we categorized models into two groups: Unaligned and Aligned, depending on the data setting. Generally,

Table 6.2 Results on MOSI and MOSEI

Model	MOSI				MOSEI				Data Setting
	MAE	Corr	Acc-2	F1-Score	MAE	Corr	Acc-2	F1-Score	Unaligned
TFN (B) [12]	0.901	0.698	−/80.8	−/80.7	0.593	0.700	−/82.5	−/82.1	Unaligned
LMF (B) [12]	0.917	0.695	−/82.5	−/82.4	0.623	0.677	−/82.0	−/82.1	Aligned
MFN	0.965	0.632	77.4/−	77.3/−	−	−	76.0/−	76.0/−	Aligned
RAVEN	0.915	0.691	78.0/−	76.6/−	0.614	0.662	79.1/−	79.5/−	Aligned
MFM (B) [12]	0.877	0.706	−/81.7	−/81.6	0.568	0.717	−/84.4	−/84.3	Aligned
MulT (B) [12]	0.861	0.711	81.5/ 84.1	80.6/ 83.9	0.58	0.703	−/82.5	−/82.3	Aligned
MISA (B) [12]	0.783	0.761	81.8/ 83.4	81.7/ 83.6	0.555	0.756	83.6/ 85.5	83.8/ 85.3	Aligned
MAG-BERT (B) [11]	0.712	0.796	84.2/ 86.1	84.1/ 86.0	−	−	84.7/−	84.5/−	Aligned
MISA (B) *	0.804	0.764	80.79/ 82.1	80.77/ 82.03	0.568	0.724	82.59/ 84.23	82.67/ 83.97	Aligned
MAG-BERT (B) *	0.731	0.789	82.54/ 84.3	82.59/ 84.3	0.539	0.753	83.79/ 85.23	83.74/ 85.08	Aligned
Self-MM (B)*	0.713	0.798	84.00/ 85.98	84.42/ 85.95	0.530	0.765	82.81/ 85.17	82.53/ 85.30	Unaligned

Models with * are reproduced under the same conditions

models that utilize aligned corpus tend to perform better. Our experimental results demonstrate that our method significantly outperforms unaligned models (TFN and LMF) in all evaluation metrics. Moreover, even when compared with aligned models, our approach still achieves competitive results. We also reproduced two of the best baselines "MISA" and "MAG-BERT" under the same conditions and found that our model surpasses them in most evaluation metrics. With the SIMS dataset only containing unaligned data, we compare the Self-MM with TFN and LMF. Additionally, we use human-annotated unimodal labels to replace the autogenerated u-labels, known as Human-MM. The experimental results are presented in Table 6.3, which illustrate that our Self-MM approach yields better results than TFN and LMF and achieves comparable performance with Human-MM. Overall, these results indicate that our model can be effectively applied to different data scenarios, resulting in significant improvements in performance.

Table 6.3 Results on SIMS

Model	MAE	Corr	Acc-2	F1-Score
TFN	0.428	0.605	79.86	80.15
LMF	0.431	0.600	79.37	78.65
Human-MM	0.408	0.647	81.32	81.73
Self-MM	0.419	0.616	80.74	80.78

6.1.4 Conclusion

In this section, we present a novel approach to learning modality-specific representations by introducing unimodal subtasks. Our method differs from previous works in that we utilize a self-supervised method to generate unimodal labels, significantly reducing the need for human annotation. Through extensive experimentation, we demonstrate the reliability and stability of the autogenerated unimodal labels. This work offers a fresh perspective on multimodal representation learning. However, we also observed that the generated audio and vision labels have limitations due to the preprocessed features. In future work, we plan to develop an end-to-end multimodal learning network and investigate the relationship between unimodal and multimodal learning.

6.2 Multimodal Sentiment Analysis Method Based on Modality Missing

6.2.1 Introduction

Enhancing the robustness of Multimodal Sentiment Analysis (MSA) against data missing has emerged as a key challenge. MSA aims to assess speaker sentiments based on language, visual, and acoustic signals. In previous research, translation-based and tensor regularization methods have been proposed for MSA with incomplete modality features. However, both approaches fail to handle random modality feature missing in nonaligned sequences. To address this issue, this section presents a transformer-based feature reconstruction network (TFR-Net) that improves model robustness against random missing modality features in nonaligned sequences. The proposed method involves utilizing intramodal and intermodal attention-based extractors to learn resilient representations for each element in modality sequences. A reconstruction module is then employed to generate the missing modality features. By supervising the model with Smooth L1 Loss between generated and complete sequences, TFR-Net is expected to learn semantic-level features corresponding to missing features. Extensive experimentation on two publicly available benchmark datasets demonstrates that our proposed model achieves excellent performance against data missing across various missing modality combinations and degrees of missingness.

6.2.2 Methodology

In this section, we present our technique for developing resilient representations against missing modalities by employing modality reconstruction. The framework is visually depicted in Fig. 6.3.

Task Setup

Our objective is to measure the sentiments in videos using incomplete multimodal signals. Each video clip consists of three sequences of low-level features with random missing values from text (t), audio (a), and visual (v), represented as $U'_t \in R^{T_t \times d_t}$, $U'_a \in R^{T_a \times d_a}$, $U'_v \in R^{T_v \times d_v}$, respectively. The proposed model takes U'_t, U'_a, and U'_v as inputs and produces a single sentimental intensity result, y^m. During the training stage, complete modality features $U_t \in R^{T_t \times d_t}$, $U_a \in R^{T_a \times d_a}$, $U_v \in R^{T_v \times d_v}$ are utilized, along with the feature missing positions, to guide the representation learning process.

Modality Feature Extraction Module

Initially, the modality feature extraction module utilizes a 1D convolutional layer to process the incomplete modality sequences, thereby ensuring that each element of the input sequences is cognizant of its neighboring elements.

$$H_m = \mathrm{Conv1\,d}\left(U'_m, k_m\right) \in R^{T_m, d}, m \in \{t, a, v\} \tag{6.19}$$

The modality feature extraction module employs convolutional kernels of size $k_{t,\,a,\,v}$ for each modality (t, a, v), with a common dimension d. The convolved sequences are then augmented with position embedding (PE) and fed into intra-modal and inter-modal transformers to capture modality dynamics for each time-step of the input sequences. To extract information from one sequence H_i to another sequence H_j, we utilize the attention mechanism in the transformer encoder structure. The transformer encoder accepts queries, keys, and values as inputs, with the queries sourced from H_i and the keys and values sourced from H_j. Thus, the transformer encoder can be represented as Transformer(H_i, H_j, H_j):

$$H'_m = H_m + \mathrm{PE}_m(T_m, d) \tag{6.20}$$

$$H_{m \to m} = \mathrm{Transformer}\left(H'_m, H'_m, H'_m\right) \in R^{T_m, d} \tag{6.21}$$

$$H_{n \to m} = \mathrm{Transformer}\left(H'_m, H'_n, H'_n\right) \in R^{T_m, d} \tag{6.22}$$

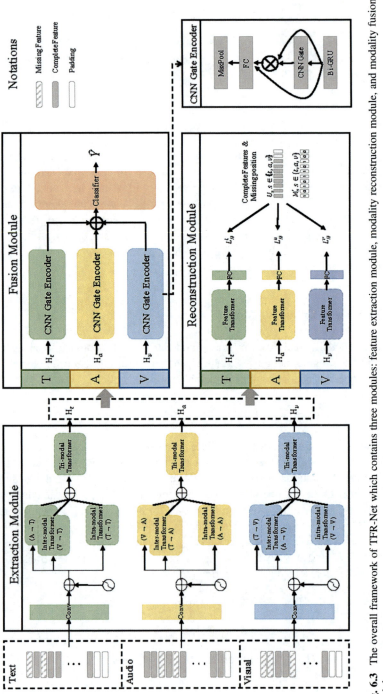

Fig. 6.3 The overall framework of TFR-Net which contains three modules: feature extraction module, modality reconstruction module, and modality fusion module

where $\text{PE}_m(T_m, d) \in R^{T_m, d}$ generates embeddings for each position index, $m \in \{t, a, v\}$, $n \in \{t, a, v\} - \{m\}$.

Ultimately, we concatenate all latent features obtained from the intra-modal and inter-modal transformers to generate the enhanced sequence features output:

$$H = \text{Concat}\left([H_{m \to m}; H_{n_1 \to m}; H_{n_2 \to m}]\right) \in R^{T_m, 3d} \tag{6.23}$$

$$H''_m = \text{Transformer}\left(H, H, H\right) \in R^{T_m, 3d} \tag{6.24}$$

where contains the variables $m \in \{t, a, v\}$ and n_1, n_2, which pertain to the other two modalities apart from m. The enhanced sequences are tasked with extracting efficient representations for the missing modality features by leveraging the complementary nature of the modalities. Furthermore, these enhanced modality sequences, which incorporate cross-modal interactions, can be seen as the outcome of fusion at the model level.

Modality Reconstruction Module

Our approach involves a Modality Reconstruction (MR) Module, which is founded on the crucial observation that reconstructing complete modality sequences from the extracted modality sequences can aid the extractor module in comprehending the semantics of the missing parts. For every modality, we begin by implementing a self-attention mechanism on the feature dimension to capture the interactions among the extracted features:

$$H^*_m = \text{Transformer}\left(H''^T_m, H''^T_m, H''_m\right)^T \in R^{T_m, 3d} \tag{6.25}$$

this equation pertains to the variables $m \in \{t, a, v\}$, where H^*_m is deemed as the transformed sequence features. Subsequently, we carry out a linear transformation that maps the extracted features back into the input spaces:

$$\hat{U}_m = W_m \cdot H^*_m + b_m \tag{6.26}$$

where involves the variables $m \in \{t, a, v\}$, with W_m and b_m representing the parameters of the linear layer.

To supervise the training process, we employ the SmoothL1Loss $(\cdot)$ function, which calculates the loss between the original and generated values for the missing elements. This generation loss, denoted as $\mathcal{L}^m_g$, helps to account for the impact of missing reconstruction:

$$\mathcal{L}_g^m = \text{SmoothL1Loss}\left(\hat{U}_m * M', U_m * M'\right) \tag{6.27}$$

we have $m \in \{t, a, v\}$ and M' represents the missing mask that discloses the positions where the input modality sequences are missing.

Fusion Module

Following the integration of complementary modality information into the incomplete modality sequences via reconstruction loss, we merge them into a consolidated vector for sentiment predictions. To encode the enhanced modality sequences $\overline{H}_m$, we adopt the proposed CNN Gate Encoder, which processes them separately.

The CNN Gate Encoder involves multiple stages. At the outset, the extracted modality sequences $\overline{H}_m$ undergo a bidirectional GRU layer, which is succeeded by the application of the tanh activation function to yield the updated representation H_m''.

$$\overline{H}_m = \text{tanh}\left(\text{BiGRU}\left(H_m''\right)\right) \tag{6.28}$$

To enhance the encoding of $\overline{H}_m$, we introduce a Convolution Gate component. This component employs a one-dimensional convolution network (CNN) that moves a convolution kernel with window size k over the input H_m'' to derive a scalar value g_i for each element in the sequences. To ensure that H_m'' and g have the same sequence length, we apply a padding strategy.

$$g = \text{sigmoid}\left(\text{Conv1d}\left(\overline{H}_m\right)\right) \tag{6.29}$$

where $m \in \{t, a, v\}$. One-dimensional convolution operation is performed on these utterances using a function called $\text{Conv1d}(\cdot)$. To eliminate irrelevant contextual information, a gate called g is used to rescale the representation $\overline{H}_m$:

$$\overline{H}_m' = \overline{H}_m \bigotimes g \tag{6.30}$$

the symbol $\bigotimes$ indicates an element-wise multiplication operation.

After obtaining the representation $\overline{H}_m'$ and the initial extracted sequences H_m'', they are combined by concatenation. The resulting sequence is then passed through a fully connected layer to adjust the dimensionality of the final word-level representation H_m^*.

$$H_m^* = \text{tanh}\left(W \cdot \text{Concat}\left[\overline{H}_m', H_m''\right] + b\right) \tag{6.31}$$

To obtain a final modality representation U_m^*, we utilize the max-pooling operation to emphasize the features in an utterance that have a greater influence. This operation selects the most significant features from the previous layers and combines them to create a condensed representation of the modality.

$$U_m^* = \text{Maxpool}\{H_m^*\} \in R^{h_m} \tag{6.32}$$

The variable h_m is used to denote the hidden dimension of the modality m:

$$U^* = \text{Concat}\left[U_t^*, U_a^*, U_v^*\right] \tag{6.33}$$

$$\hat{y} = W_1 \cdot \text{LeakyReLU}\left(W_2 \cdot \text{BN}(U^*) + b_2\right) + b_1 \tag{6.34}$$

BatchNorm is a mathematical operation that is applied to a batch of data, and LeakyReLu is the activation function used in this context.

Model Training

To train our model for sentiment intensity prediction, we use L1 Loss as the primary optimization objective. In addition to this, we also include a reconstruction loss, denoted as $L_g^m, m \in \{t, a, v\}$, in the overall learning process. The goal is to minimize the combined loss from both L1 Loss and reconstruction loss:

$$\mathcal{L}_{\text{gen}} = \sum_{m \in \{t, a, v\}} \lambda_m \cdot L_g^m \tag{6.35}$$

$$\mathcal{L} = \frac{1}{N} \sum_{i}^{N} \left(\left|\hat{y}\,i - y^i\right|\right) + \mathcal{L}_{\text{gen}} \tag{6.36}$$

the weights $\lambda_m, m \in \{t, a, v\}$ are used to assign different levels of importance to each modality reconstruction loss L_g^m, which collectively contribute to the overall loss $\mathcal{L}$. Each of these individual loss components is responsible for learning the representations within its respective modality subspace.

6.2.3   Experiments

This section outlines our experimental methodology for assessing the resilience of our model to random missing features across different modalities.

Datasets

This study involves performing experiments on two publicly available multimodal sentiment analysis datasets, namely MOSI [4] and SIMS [6]. Table 6.4 presents some basic statistics for each dataset. Below, we provide a brief overview of these datasets.

1. **MOSI**: One of the most widely used benchmark datasets for multimodal sentiment analysis (MSA) is the CMU-MOSI dataset. It contains 2199 brief monologue video clips that are extracted from 93 movie review videos on YouTube. Each sample is assigned a sentiment score by human annotators, ranging from −3 (strongly negative) to 3 (strongly positive).
2. **SIMS**: The SIMS dataset is a unique benchmark for multimodal sentiment analysis (MSA) in the Chinese language, featuring finely grained annotations of modality. The dataset comprises 2,281 high-quality video clips that are meticulously selected from a range of movies, TV series, and variety shows, showcasing spontaneous expressions, diverse head poses, occlusions, and illuminations. Each sample is assigned a sentiment score by human annotators, ranging from −1 (strongly negative) to 1 (strongly positive).

Feature Extraction

To process information from videos across the three modalities, we follow the following approach.

1. **Text Modality**: To extract features from the text modality in both the MOSI and SIMS datasets, we use a pretrained BERT [13] model as a feature extractor. This model is used to encode transcribed word sequences into text modality features, with a dimensionality of d_t equal to 768.
2. **Audio Modality**: To extract audio features, we use different acoustic frameworks for the MOSI and SIMS datasets. Specifically, we utilize the COVAREP acoustic framework [14] for the MOSI dataset and the LibROSA [15] framework for the SIMS dataset. The resulting feature dimensions are d_a equal to 5 for MOSI and 33 for SIMS dataset.
3. **Visual Modality**: To extract facial expression features for the MOSI dataset, we use Facet1. On the other hand, for the SIMS dataset, we use the MTCNN face detection algorithm [16] to detect and align faces, followed by the extraction of facial features using the Multi Comp OpenFace2.0 toolkit [17]. The resulting feature dimensions are d_v equal to 20 for MOSI and 709 for SIMS.

Table 6.4 Dataset statistics for benchmark MSA dataset in format negative/neutral/positive

Dataset	# Train	# Valid	# Test	# All
MOSI	552/53/679	92/13/124	379/30/277	2199
SIMS	742/207/419	248/69/139	248/69/140	2281

Baselines

To assess the performance of our proposed model, we conduct experiments on three baseline methods that are capable of processing unaligned multimodal datasets.

1. **TFN**: The Tensor Fusion Network (TFN) [1] employs a tensor fusion layer that utilizes the cartesian product to create a feature vector. This enables the fusion of information from all three modalities to predict the sentiment.
2. **MulT**: The Multimodal transformer (MulT) [2] utilizes a crossmodal attention mechanism to learn the interrelationships between different modalities. This enables the model to capture the interactions between modalities and results in improved performance on unaligned multimodal datasets.
3. **MISA**: The method used in this study learns both modality-invariant and modality-specific representations [12] by projecting each modality of samples into two subspaces. This feature extraction technique is highly efficient and has been shown to significantly improve the performance of models in multimodal sentiment analysis tasks.

Experimental Settings

To evaluate the robustness of our proposed model, we created multimodal datasets with missing values by randomly replacing some values in the sequence with [UNK] in the text modality and zero padding vectors in other modalities. The proportion of missing values in each modality is specified in advance and kept consistent across the training, validation, and test datasets.

To optimize our proposed model, we tuned various hyperparameters, including the convolution kernel size, attention dropout, number of heads in transformers, the dimension of feature vectors for fusion, and weights of generative loss for each modality. These parameters were optimized separately for each dataset on the validation set. For training, we utilized the Adam optimizer with a learning rate of 0.002 for the MOSI dataset and 0.001 for the SIMS dataset. The evaluation results are based on the average of three experiments using different random seeds for both datasets.

Evaluation Metrics

To conduct a comprehensive comparison with the baseline methods, we evaluated the performance of our proposed model and other methods using various metrics on the MOSI and SIMS test sets. We recorded binary classification accuracy (Acc-2), five classification accuracy (Acc-5), Mean Absolute Error (MAE), and Pearson Correlation coefficient (Corr) as the missing rate increased. We followed the approach used in recent studies [2, 12] to calculate the binary classification accuracy

(Acc-2), which assigns negative and positive classes for sentiment scores <0 and >0, respectively.

Additionally, we calculated the Area Under Indicators Line Chart (AUILC) for each metric sequence to evaluate the overall performance of the methods in dealing with incomplete modality inputs quantitatively. The Area Under Indicators Line Chart (AUILC) is a quantitative measure used to assess the performance of the different models in handling incomplete modality inputs. It is computed based on a sequence of model evaluation results $X = \{x_0, x_1, \cdots, x_t\}$ obtained at increasing missing rates $\{r_0, r_1, \cdots, r_t\}$.

$$\text{AUILC}_X = \sum_{i=0}^{t-1} \frac{(x_i + x_{i+1})}{2} \cdot (r_{i+1} - r_i) \tag{6.37}$$

For all the metrics mentioned above, a higher value indicates better performance, except for Mean Absolute Error (MAE), where a lower value indicates stronger performance.

Results and Discussion

In this section, we provide a detailed analysis and discussion of the results obtained from our experiments.

We begin by analyzing the robustness of the TFR-Net model to increasing levels of random modality missing rates. For the experiments, we introduce the same missing rate in each modality during both the training and testing periods, with the missing rate parameterized by missing rate $\in \{0.0, 0.1, \cdots, 1.0\}$. To simulate the missing values, we use a random drop strategy, where each entry is independently dropped with probability $p \in$ missing rate.

Question 1

How does the performance of TFR-Net compare to that of existing approaches for multimodal sentiment analysis?

To evaluate the effectiveness of the TFR-Net model in comparison to existing approaches for multimodal sentiment analysis, we first present the performance curves for the model. As shown in Fig. 6.4, on the MOSI dataset, TFR-Net

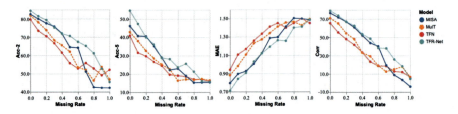

Fig. 6.4 Metrics curves of various missing rates on MOSI dataset

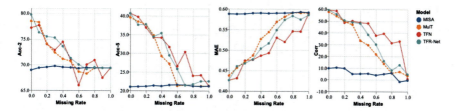

Fig. 6.5 Metrics curves of various missing rates on SIMS dataset

Table 6.5 AUILC results comparison with baseline models on MOSI and SIMS dataset

Models	MOSI				SIMS			
	Acc-2 (↑)	Acc-5 (↑)	MAE (↓)	Corr (↑)	Acc-2 (↑)	Acc-5 (↑)	MAE (↓)	Corr (↑)
TFN	0.604	0.233	1.327	0.300	0.373	**0.181**	**0.233**	**0.259**
MulT	0.618	0.244	1.288	0.334	0.370	0.173	0.244	0.227
MISA	0.632	0.271	1.209	0.403	0.347	0.106	0.294	0.038
TFR-net	**0.690**	**0.304**	**1.155**	**0.467**	**0.377**	0.180	0.237	0.253

outperforms the baseline approaches on most evaluation metrics for all missing rates $p \in \{0.0, 0.1 \cdots, 1.0\}$. Similarly, on the SIMS dataset, as shown in Fig. 6.5, TFR-Net achieves better performance under low missing rates ($p \in \{0.0, 0.1, \cdots, 0.5\}$). However, under higher missing rates ($p \in \{0.6, 0.7, \cdots, 1.0\}$), all models perform similarly and eventually converge to a stable value. We attribute this phenomenon to the label bias present in the dataset.

To further evaluate the performance of the proposed model, we compute the AUILC value for the whole interval $p \in \{0.0, 0.1 \cdots, 1.0\}$ on the MOSI dataset and the AUILC value for the partial interval ($p \in \{0.0, 0.1, \cdots, 0.5\}$) on the SIMS dataset. The results are presented in Table 6.5. The quantitative results further confirm the robustness of the proposed TFR-Net model for various modality missing reason both datasets. In addition to the performance curves and AUILC values, we also present the quantitative results in Table 6.5. The results for the MOSI dataset were calculated with the whole missing rate interval $p \in \{0.0, 0.1 \cdots, 1.0\}$, while the results for the SIMS dataset were calculated with the partial missing rate interval ($p \in \{0.0, 0.1, \cdots, 0.5\}$). The results demonstrate the effectiveness of the TFR-Net model in handling missing values in multimodal datasets, outperforming the baseline approaches across a range of evaluation metrics. Overall, these results support the claim that TFR-Net is a robust and effective model for multimodal sentiment analysis in the presence of missing values.

In our subsequent experiment, we investigate the ability of the TFR-Net model to handle different combinations of missing modalities. To accomplish this, we performed experiments on the MOSI dataset using the TFR-Net model, where we completely dropped different combinations of modalities with a missing rate of $p = 1.0$.

Table 6.6 TFR-Net results for different modality missing combinations

Test Input	MOSI			
	Acc-2($\uparrow$)	Acc-5($\uparrow$)	MAE($\downarrow$)	Corr($\uparrow$)
{a}	55.15	16.57	1.419	0.214
{v}	60.11	17.49	1.381	0.164
{t}	83.49	50.14	0.786	0.778
{a, v}	62.65	19.05	1.334	0.231
{t, a}	83.99	52.92	**0.731**	**0.788**
{t, v}	82.62	49.37	0.772	0.778
{t, a, v}	**84.10**	**54.66**	0.754	0.783

Question 2

How does the performance of TFR-Net vary under different combinations of missing modalities during testing?

Table 6.6 presents the experimental results obtained from our study. All reported results are the average of three groups of seeds. In the experiments where only a single modality was available, we found that TFR-Net maintained comparable performance when the text modality was available, while its performance was significantly inferior when either the audio or visual modality was available. In the experiments where two modalities were available, we observed that TFR-Net performed best when text modality was combined with visual modality, achieving even better MAE and Corr compared to the experiments with all three modalities. Based on the above results, we can conclude that the text modality contains more semantics and plays a critical role in reconstructing missing semantics and predicting sentiment. However, it is challenging for the model to reconstruct the semantics present in the text modality with audio and visual features input.

6.2.4 Conclusion

In this section, our focus has been on improving the robustness of models against the incompleteness of modalities for the Multimodal Sentiment Analysis (MSA) task. To achieve this, we have proposed a transformer-based feature reconstruction network (TFR-Net), which is a flexible framework capable of handling the incompleteness of nonaligned features in various modality combinations and degrees. The core of our proposed TFR-Net is the feature reconstruction module, which guides the extractor to acquire semantics of the missing modalities features. Our experimental results on two benchmark MSA datasets show that our model performs well in the presence of incompleteness of nonaligned features in various modalities and degrees. However, we also found that the current model performance is constrained by the label bias problem, which we plan to address in future work.

6.3 Summary

In this chapter, we introduce a novel approach to learning modality-specific representations using unimodal subtasks. Unlike previous works, they employ a self-supervised method to generate unimodal labels, reducing the reliance on human annotation. Extensive experimentation confirms the reliability and stability of the autogenerated unimodal labels, providing a fresh perspective on multimodal representation learning. However, limitations are observed in the generated audio and vision labels due to preprocessed features. Future work aims to develop an end-to-end multimodal learning network and investigate the relationship between unimodal and multimodal learning. This chapter also discusses efforts to improve the robustness of models for Multimodal Sentiment Analysis (MSA) by addressing the incompleteness of modalities. The proposed framework, TFR-Net, utilizes a transformer-based feature reconstruction module to handle nonaligned features in various modality combinations and degrees. Experimental results on benchmark MSA datasets demonstrate the model's effectiveness in the presence of incomplete nonaligned features. However, the model's performance is constrained by the label bias problem, which will be addressed in future work.

References

1. Zadeh A, Chen M, Poria S, et al (2017) Tensor fusion network for multimodal sentiment analysis. Proceedings of the 2017 Association for Computational Linguistics Conference on Empirical Methods in Natural Language Processing, 1103–1114
2. Tsai YHH, Bai S, Liang PP, et al (2019) Multimodal transformer for unaligned multimodal language sequences. Proceedings of the 57th Annual Meeting of the Association for Computational Linguistics, 6558–6569
3. Poria S, Hazarika D, Majumder N et al (2020) Beneath the tip of the iceberg: current challenges and new directions in sentiment analysis research. IEEE Trans Affect Comput:1–29
4. Zadeh A, Zellers R, Pincus E, et al (2016) Mosi: multimodal corpus of sentiment intensity and subjectivity analysis in online opinion videos. arXiv preprint arXiv:1606.06259
5. Zadeh AAB, Liang PP, Poria S, et al (2018) Multimodal language analysis in the wild: Cmu-mosei dataset and interpretable dynamic fusion graph. Proceedings of the 56th Annual Meeting of the Association for Computational Linguistics, (1): 2236–2246
6. Yu W, Xu H, Meng F, et al (2020) Ch-sims: A chinese multimodal sentiment analysis dataset with fine-grained annotation of modality. Proceedings of the 58th Annual Meeting of the Association for Computational Linguistics, 3718–3727
7. Liu Z, Shen Y, Lakshminarasimhan VB, et al (2018) Efficient low-rank multimodal fusion with modality-specific factors. Proceedings of the 56th Annual Meeting of the Association for Computational Linguistics, (1):2247–2256
8. Zadeh A, Liang PP, Mazumder N, et al (2018) Memory fusion network for multi-view sequential learning. Proceedings of the 32nd Association for the Advancement of Artificial Intelligence Conference on Artificial Intelligence, 5634–5641
9. Tsai YHH, Liang PP, Zadeh A, et al (2019) Learning factorized multimodal representations. Proceedings of the 2019 International Conference on Representation Learning.

10. Wang Y, Shen Y, Liu Z et al (2019) Words can shift: dynamically adjusting word representations using nonverbal behaviors. Proc Assoc Advanc Artific Intellig Conf Artific Intellig 33(1): 7216–7223
11. Rahman W, Hasan MK, Lee S, et al (2020) Integrating multimodal information in large pretrained transformers. Proceedings of the 58th Annual Meeting of the Association for Computational Linguistics, 2359–2369
12. Hazarika D, Zimmermann R, Poria S (2020) Misa: Modality-invariant and-specific representations for multimodal sentiment analysis. Proceedings of the 28th Association for Computing Machinery International Conference on Multimedia, 1122–1131
13. Devlin J, Chang M, Lee K, et al (2019) Bert: Pre-training of deep bidirectional transformers for language understanding. Proceedings of the 2019 Conference of the North American Chapter of the Association for Computational Linguistics: Human Language Technologies, (1):4171–4186
14. Degottex G, Kane J, Drugman T, et al (2014) COVAREP—A collaborative voice analysis repository for speech technologies. Proceedings of the 2014 Institute of Electrical and Electronics Engineers international conference on acoustics, speech and signal processing, 960–964
15. Brian McFee, Colin Raffel, Dawen Liang, et al. (2015) Librosa: Audio and music signal analysis in python. Proceedings of the 14th python in science conference, 18–25
16. Zhang K, Zhang Z, Li Z et al (2016) Joint face detection and alignment using multitask cascaded convolutional networks. IEEE Sig Process Lett 23(10):1499–1503
17. Baltrusaitis T, Zadeh A, Lim YC, et al (2018) OpenFace 2.0: Facial behavior analysis toolkit. Proceedings of the 13th Institute of Electrical and Electronics Engineering International Conference on Automatic Face & Gesture Recognition, 59–66

Chapter 7
Multimodal Sentiment Analysis Platform and Application

Abstract This chapter introduces the need for an integrated platform specifically designed for Multimodal Sentiment Analysis (MSA) tasks. To address this gap, the M-SENA platform is introduced as an open-source tool aimed at facilitating advanced research in MSA. The platform is built on the principles of flexibility, reliability, and intuitive usability, offering researchers and practitioners a range of toolkits, benchmarks, and demonstrations to drive innovation in the field. This chapter provides an overview of the M-SENA platform's architecture and highlights the key features of its core modules. It also presents reliable baseline results of different modality features and MSA benchmarks to demonstrate the platform's effectiveness. The evaluation and analysis tools provided by M-SENA are utilized to assess model performance, visualize intermediate representations, conduct instance tests, and evaluate model generalization ability. Furthermore, the preface acknowledges the importance of model robustness against modality noise in real-world applications. The introduction of Robust-MSA, an interactive platform, is described to help researchers visualize the impact of modality noise on models and explore defense methods to improve robustness. The goal of Robust-MSA is to provide researchers with a better understanding of model performance with imperfect real-world data and identify areas for improvement.

7.1 An Integrated Platform for Multimodal Sentiment Analysis

7.1.1 Introduction

While previous works have demonstrated significant improvements on benchmark datasets [1–4], there is currently a lack of integrated platforms specifically designed for Multimodal Sentiment Analysis (MSA) tasks. To address this gap, we are proud to announce the release of the first comprehensive MSA platform.

M-SENA is an open-source platform designed to facilitate advanced research in the field of Multimodal Sentiment Analysis (MSA). Our platform is built on the principles of flexibility, reliability, and intuitive usability, providing researchers and practitioners with a range of toolkits, benchmarks, and demonstrations to drive innovation in the field. The platform features a fully modular video sentiment analysis framework, which includes modules for data management, feature extraction, model training, and result analysis. In this section, we provide an overview of the M-SENA platform's overall architecture and highlight the key features of its core modules. Additionally, we present reliable baseline results of different modality features and MSA benchmarks to demonstrate the platform's effectiveness. To further evaluate the performance of the models, we utilize the model evaluation and analysis tools provided by M-SENA. These tools enable us to visualize intermediate representations, conduct on-the-fly instance tests, and evaluate the generalization ability of the models.

7.1.2   Platform Architecture

The M-SENA platform is a comprehensive tool for Multimodal Sentiment Analysis (MSA) that offers convenient data access, customized feature extraction, a unified model training pipeline, and comprehensive model evaluation. Researchers can access the platform through a graphical web interface or Python packages, which provide all the features mentioned above.

The platform currently supports three popular MSA datasets across two languages, seven feature extraction backbones, and fourteen benchmark MSA models. Figure 7.1 provides an overview of the M-SENA platform's architecture. In the following sections, we describe the features of each module in Fig. 7.1 in detail.

Data Management Module

The data management module in M-SENA is a powerful tool that simplifies the process of accessing multimedia data on servers. In addition to providing access to existing benchmark datasets, this module also allows researchers to create and manage their own datasets, making it a versatile tool for a range of research applications. With the data management module, users can easily import and preprocess multimodal data, and organize it into a format suitable for MSA tasks. This can include functions like data augmentation, labeling, and filtering, among others. Additionally, the module offers users the flexibility to customize their datasets, adding their own data or modifying existing datasets to meet their specific research needs.

1. **Benchmark Datasets**: M-SENA currently provides support for three benchmark MSA datasets in English and Chinese, including CMU-MOSI [5], CMU-MOSEI [6], and CH-SIMS [4].

   With the platform's built-in dataset management tools, users can easily filter and view raw videos without the need to download them to their local

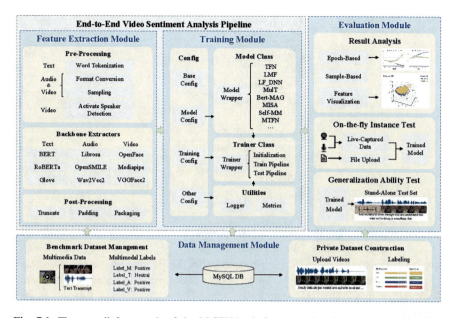

Fig. 7.1 The overall framework of the M-SENA platform contains four main modules: data management module, feature extraction module, model training module and model evaluation module

environment. This feature streamlines the data access process and enables researchers to work more efficiently with large and complex multimedia datasets.

2. **Building Private Datasets**: In addition to supporting benchmark MSA datasets, the M-SENA platform also offers a graphical interface that enables researchers to construct their own datasets using uploaded videos. Based on the literature [4], M-SENA supports both unimodal and multimodal sentiment labeling.

   Users can label sentiment in different modalities of the uploaded videos, and these labeled datasets can be used directly for model training and evaluation on the platform. This feature provides researchers with a powerful tool for creating custom datasets tailored to their specific research needs.

Feature Extraction Module

Extracting emotion-bearing features from different modalities remains a significant challenge for Multimodal Sentiment Analysis (MSA) tasks. To address this challenge, the M-SENA platform offers users access to seven commonly used feature extraction tools, enabling effective extraction of modality features for MSA.

The platform provides a unified Python API and graphical interface for these feature extraction tools, making it easy for researchers to use and customize these tools as needed. Table 7.1 provides an overview of the supported features for each modality, which are described in detail below.

Table 7.1 Some of the supported features in M-SENA

Acoustic feature sets	
ComParE_2016 [7]	Static (HSFs)
eGeMAPS [8]	Static (LLDs)
wav2vec2.0 [9]	Learnable
Visual feature sets	
Facial landmarks [10]	Static
Eyes gaze [11]	Static
Action unit [12]	Static
Textual feature sets	
GloVe6B [13]	Static
BERT [14]	Learnable
RoBerta [15]	Learnable

1. **Acoustic Modality**: Previous research has demonstrated the effectiveness of various acoustic features for emotion recognition [16, 17]. These features can be divided into two types: low-level descriptors (LLDs) and high-level statistical functions (HSFs). LLDs, such as prosodies and spectral domain features, are calculated on a frame-by-frame basis, while HSFs are calculated on an entire utterance level.

   In addition to hand-crafted features, the M-SENA platform also provides a pretrained acoustic model, wav2vec2.0 [9], which can be used as a learnable feature extractor. Researchers can also design and build their own customized acoustic features using the provided Librosa extractor.

2. **Visual Modality**: Facial landmarks, eye gaze, and facial action units are commonly used visual features in existing MSA research. The M-SENA platform offers researchers the flexibility to extract visual feature combinations using OpenFace and MediaPipe extractors.

   These extractors provide a range of tools for extracting various visual features, including facial landmarks, eye gaze, and facial action units. M-SENA allows users to customize their feature extraction pipelines based on their specific research needs, enabling them to effectively extract emotion-bearing features from visual modalities for use in MSA tasks.

3. **Text Modality**: In comparison to acoustic and visual features, semantic text embeddings have become more mature with the rapid development of pretrained language models [18]. The M-SENA platform provides support for a range of popular pretrained language models as textual feature extractors, following previous works [1, 3, 19]

   M-SENA supports GloVe6B [13], pretrained BERT [14], and pretrained RoBERTa [20] as textual feature extractors. These models offer a range of powerful tools for extracting semantic information from text data, enabling researchers to effectively capture sentiment cues from text modalities for use in MSA tasks.

Model Training Module

The M-SENA platform provides a unified training module that currently integrates 14 MSA benchmarks. These benchmarks include a range of different methods, such as tensor fusion, modality factorization, word-level fusion, multiview learning, and other MSA methods.

By integrating these benchmarks, the M-SENA platform enables researchers to easily compare the performance of different MSA models and evaluate their effectiveness. Moreover, the platform will continue to follow advanced MSA benchmarks and provide reliable benchmark results for future MSA research.

Result Analysis Module

The M-SENA platform offers a range of comprehensive model evaluation tools to support researchers in analyzing the performance of their MSA models. These tools include intermediate result visualization, on-the-fly instance testing, and generalization ability testing.

Intermediate result visualization enables researchers to visualize the intermediate representations of their models, providing valuable insights into the model's decision-making process. On-the-fly instance testing allows researchers to test their models on new, unseen instances, enabling them to evaluate the model's performance in real-world scenarios. Generalization ability testing enables researchers to evaluate the model's ability to generalize to new, unseen datasets, providing important insights into the model's overall effectiveness. These model evaluation tools provide researchers with a powerful set of tools for analyzing the performance of their MSA models, enabling them to identify areas for improvement and develop more effective models.

1. **Intermediate Result Visualization**: Multimodal representation discrimination is a crucial metric for evaluating the effectiveness of different fusion methods. The M-SENA platform provides researchers with a range of tools for evaluating the performance of their models, including the recording and illustration of final multimodal fusion results after decomposition with Principal Component Analysis (PCA).

   Additionally, M-SENA offers training loss, binary accuracy, and F1 score curves for detailed analysis of model performance. These tools enable researchers to gain a detailed understanding of the performance of their models, identify areas for improvement, and develop more effective MSA models.
2. **Live Demo Module**: The M-SENA platform aims to bridge the gap between MSA research and real-world video sentiment analysis scenarios by providing a live demo module. This module enables researchers to perform on-the-fly instance tests, validating the effectiveness and robustness of their selected MSA model by uploading or live-feeding videos to the platform.

Table 7.2 Statistics of the generalization ability test dataset, where "en" represents "English", "ch" represents "Chinese"

| | | Scenarios | |
Types	Films(TV)	Variety show	Life(Vlog)
Easy	10 (en:4 ch:6)	8 (en:4 ch:4)	8 (en:4 ch:4)
Common	9 (en:4 ch:5)	11 (en:6 ch:5)	8 (en:4 ch:4)
Difficult	9 (en:4 ch:5)	9 (en:5 ch:4)	8 (en:4 ch:4)
Noise	9 (en:4 ch:5)	8 (en:4 ch:4)	7 (en:2 ch:5)
Missing	9 (en:4 ch:5)	9 (en:5 ch:4)	7 (en:3 ch:4)

3. **Generalization Ability Test**: Real-world scenarios can be more complex than the test sets provided in benchmark MSA datasets. To ensure MSA models are effective in these scenarios, they must be both robust against modality noise and perform well on the test set.

To address this challenge, the M-SENA platform provides a generalization ability test dataset consisting of 68 Chinese and 61 English samples. This dataset aims to simulate a diverse range of complex real-world scenarios, driven by the demand from real-world applications and observations. The dataset's statistics are presented in Table 7.2. The generalization ability test dataset provided by the M-SENA platform consists of three scenarios and five instance types, designed to simulate a diverse range of real-world scenarios. The three scenarios include films, variety shows, and user-uploaded vlogs, while the five instance types include easy, common, and difficult samples, in addition to samples with modality noise and missing modalities.

7.1.3   Experiments

This section presents the results of experiments conducted using the M-SENA platform.

Feature Selection Comparison

In the upcoming experiments, we compare the performance of various feature sets for multimodal sentiment analysis using the M-SENA platform. By default, we use BERT [T1] for text modality, eGeMAPS (LLDs) [A1] for acoustic modality, and Action Unit [V1] for visual modality.

We compare these default feature sets with six other feature sets, including GloVe6B [T2] and RoBerta [T3] for text modality, customized acoustic features [A2] (including 20 dimensional MFCC, 12 dimensional CQT, and 1 dimensional f0), and wav2vec2.0 features [A3] for acoustic modality, and facial landmarks [V2], facial landmarks and action units [V3] for visual modality. Additionally, we report the performance of models using the modality features provided in CMU-MultimodalSDK.

Table 7.3 Results for feature selection

	TFN		GMFN		MISA		Bert-MAG	
Feature combinations	Acc-2 (%)	F1 (%)	Acc-2 (%)	F1 (%)	Acc-2 (%)	F1 (%)	Acc-2 (%)	F1 (%)
CMU-SDK†	78.02	78.09	76.98	77.06	82.96	82.98	83.41	83.47
[T1]-[A1]-[V1]	77.41	77.47	77.77	77.84	83.78	83.80	83.38	83.43
[T2]-[A1]-[V1]	70.40	70.51	71.40	71.54	75.22	75.68	–	–
[T3]-[A1]-[V1]	80.85	80.79	80.21	80.15	79.57	79.67	–	–
[T1]-[A2]-[V1]	76.80	76.82	78.02	78.03	83.72	83.72	82.96	83.04
[T1]-[A3]-[V1]	77.19	77.23	78.44	78.45	82.16	82.23	83.57	83.58
[T1]-[A1]-[V2]	77.38	77.48	78.81	78.71	83.2	83.14	82.13	82.20
[T1]-[A1]-[V3]	76.74	76.81	78.23	78.24	84.06	84.08	83.69	83.75

By evaluating the performance of these different feature sets, we can gain valuable insights into the most effective and robust feature sets for multimodal sentiment analysis. This will enable us to develop more effective models that perform well in a range of real-world scenarios.

Table 7.3 presents the results of our feature selection experiments, where we evaluated the performance of different feature sets for multimodal sentiment analysis using the M-SENA platform. The table shows the results for three modalities: text, acoustic, and visual.

For text modality, we evaluated the performance of BERT [T1], GloVe6B [T2], and RoBerta [T3]. For acoustic modality, we evaluated the performance of eGeMAPS [A1], customized features including 20-dim MFCC, 12-dim CQT, and f0 [A2], and wav2vec2.0 [A3]. For visual modality, we evaluated the performance of action units [V1], landmarks [V2], and both landmarks and action units [V3]. Additionally, we evaluated the performance of models using the modality features provided in the modified CMU-SDK features with BERT for text [CMU-SDK†] and Bert-MAG (designed upon the Bert backbone) for text.

The results show that in most cases, using appropriate features instead of the original features in CMU-MultimodalSDK can improve model performance. For text modality, RoBerta feature performs best for TFN and GMFN models, while Bert feature performs best for MISA model. For acoustic modality, wav2vec2.0 embeddings (without finetune) perform best for GMFNand Bert-MAG models. However, literature suggests that finetuning wav2vec2.0 can further improve model performance, providing more effective acoustic features for future MSA research. For visual modality, the combination of facial landmarks and action units achieves the overall best result, revealing the effectiveness of both landmarks and action units for sentiment classification.

Table 7.4 Experiment results for MSA benchmark comparison

Model	MOSI				MOSEI				SIMS			
	Acc-2	F1	MAE	Corr	Acc-2	F1	MAE	Corr	Acc-2	F1	MAE	Corr
LF_DNN	79.39	79.45	0.945	0.675	82.78	82.38	0.558	0.731	76.68	76.48	0.446	0.567
EF_LSTM	77.35	77.43	0.995	0.644	81.23	81.02	0.588	0.695	69.37	56.82	0.591	0.380
TFN	78.02	78.09	0.971	0.652	82.23	81.47	0.573	0.718	77.07	76.94	0.437	0.582
LMF	78.60	78.61	0.934	0.663	83.83	83.68	0.562	0.735	77.42	77.35	0.438	0.578
MFN	78.78	78.71	0.938	0.665	83.30	83.23	0.570	0.720	78.55	78.23	0.442	0.575
GMFN	76.98	77.06	0.986	0.642	83.48	83.23	0.575	0.713	78.77	78.21	0.445	0.578
MFM	78.63	78.63	0.958	0.649	83.49	83.29	0.581	0.721	75.06	75.58	0.477	0.525
MulT	80.21	80.22	0.912	0.695	84.63	84.52	0.559	0.733	78.56	79.66	0.453	0.564
MISA	82.96	82.98	0.761	0.772	84.79	84.73	0.548	0.759	76.54	76.59	0.447	0.563
BERT_MAG	83.41	83.47	0.761	0.776	84.87	84.85	0.539	0.764	74.44	71.75	0.492	0.399
MLF_DNN	–	–	–	–	–	–	–	–	80.44	80.28	0.396	0.665
MTFN	–	–	–	–	–	–	–	–	81.09	81.01	0.395	0.666
MLMF	–	–	–	–	–	–	–	–	79.34	79.07	0.409	0.639
Self_MM	84.30	84.31	0.720	0.793	84.06	84.12	0.531	–	80.04	80.44	0.425	0.595

MSA Benchmark Comparison

Table 7.4 presents the results of our benchmark experiments using the M-SENA platform. All models utilized the Bert embedding and the provided acoustic and visual features in CMU-MultimodalSDK. However, to improve the models' performance, we used Bert as text embeddings while using the original acoustic and visual features provided in CMU MultimodalSDK.

Due to the requirement of unimodal labels, we tested multitask models, including MLF_DNN, MTFN, and MLMF, on SIMS only.

The M-SENA platform not only provides researchers with reliable benchmark results but also offers a convenient approach to reproduce the benchmarks. The platform provides both GUI and Python API, allowing researchers to easily evaluate and compare the performance of different MSA models.

7.1.4   Model Analysis Demonstration

In this section, we present the results of our model analysis using the M-SENA platform. The platform offers a range of tools and resources for evaluating and analyzing the performance of different MSA models.

Intermediate Result Analysis

The M-SENA platform includes an intermediate result analysis submodule that is designed to monitor and visualize the training process of MSA models.

For example, Fig. 7.2 shows the training process of a TFN model on the MOSI dataset. The epoch results of binary accuracy, F1-score, and loss value are plotted to provide users with a clear picture of the model's performance during training.

In addition, the learned multimodal fusion representations are illustrated in an interactive 3D Figure, providing users with a better understanding of the multimodal feature representations and the fusion process. For models containing explicit unimodal representations, unimodal representations of text, acoustic, and visual are also shown.

On-the-Fly Instance Analysis

The M-SENA platform provides researchers with a valuable resource for validating their proposed MSA approaches using uploaded or live-recorded instances.

An example of the live demonstration is presented in Fig. 7.3, which showcases the platform's ability to provide real-time modality feature visualization along with the model prediction results. This feature enables researchers to gain a better

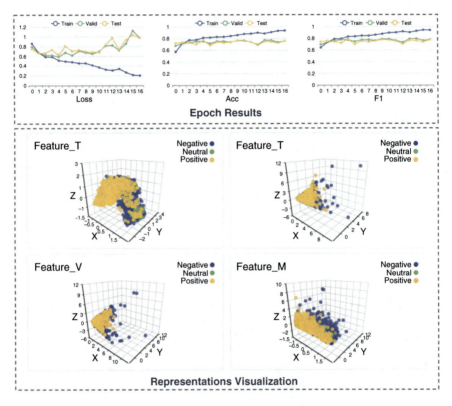

Fig. 7.2 Intermediate result analysis for TFN model trained on MOSI dataset

understanding of how their models are performing in real time and identify areas for improvement.

In addition to model prediction results, the platform also provides feature visualization, including short-time Fourier transform (STFT) for acoustic modality and facial landmarks, eye gaze, and head poses for visual modality. This feature enables researchers to gain a deeper understanding of the underlying multimodal features that are driving their model's performance.

Generalization Ability Analysis

We used the model trained on the MOSI dataset with [T1]–[A1]-[V3] modality features to conduct a generalization ability test using the M-SENA platform. The experimental results are reported in Table 7.5, where binary accuracy and F1 scores are used to evaluate the effectiveness and robustness of the model.

The results indicate that all models experienced a performance gap between the original test set and real-world scenarios, particularly for instances with noisy or missing modalities. Furthermore, the results suggest that noisy instances are

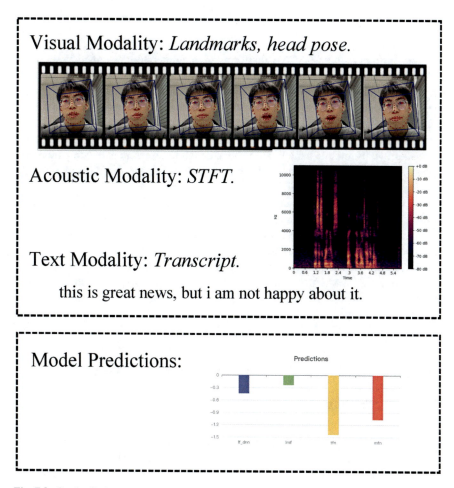

Fig. 7.3 On-the-fly instance test example

Table 7.5 Results for English generalization ability test

Types	TFN Acc-2/F1	GMFN Acc-2/F1	MISA Acc-2/F1	Bert-MAG Acc-2/F1
Easy	83.3/84.4	75.0/76.1	75.0/76.7	66.7/66.7
Common	71.4/74.5	85.7/82.3	71.4/75.8	78.6/78.6
Difficult	69.2/69.2	61.5/60.5	53.9/54.4	84.6/84.6
Noise	60.0/50.5	50.0/44.9	50.0/35.7	60.0/51.7
Missing	63.6/60.6	81.8/77.8	63.6/60.6	63.6/61.5
Avg	70.0/68.4	71.7/69.3	63.3/62.4	71.7/69.7

typically more challenging for MSA models than instances with missing modalities, highlighting that noisy modality features can be worse than no modality feature at all.

7.1.5 Conclusion

This work introduces the M-SENA platform, an integrated tool designed to support MSA researchers at every stage of the model development process. The platform offers step-by-step recipes for data management, feature extraction, model training, and model analysis, providing researchers with a comprehensive end-to-end solution for developing and evaluating MSA models.

In addition to offering comprehensive evaluation tools and resources, the platform also provides reliable benchmark results, enabling researchers to compare their models' performance against state-of-the-art benchmarks in a range of real-world scenarios.

We also provide a series of user-friendly visualization and demonstration tools, including intermediate representation visualization, on-the-fly instance test, and generalization ability test. These tools enable researchers to gain a better understanding of their models and identify areas for improvement.

In the future, we will continue to update the M-SENA platform to keep pace with the latest advances in MSA research. This will include adding new benchmarks and evaluation tools, as well as incorporating new features and models as they become available.

7.2 Robust Multimodal Sentiment Analysis Platform

7.2.1 Introduction

Ensuring model robustness against potential modality noise is a critical step toward adapting multimodal models to real-world applications. As such, researchers have increasingly focused on improving model robustness in recent years. In the context of Multimodal Sentiment Analysis (MSA), there is a debate on whether multimodal models are more effective against noisy features than unimodal ones.

To provide a more intuitive illustration and in-depth analysis of these concerns, we introduce Robust-MSA, an interactive platform designed to help researchers visualize the impact of modality noise on their models and explore simple defense methods to improve model robustness.

Robust-MSA enables researchers to gain a better understanding of how their models perform with imperfect real-world data and identify areas for improvement.

7.2.2 Demonstrating Robust-MSA

Robust-MSA is a user-friendly platform that takes user-generated videos as input for sentiment analysis. Upon uploading a video, the platform automatically proceeds with speech recognition. However, manual revision of the generated transcript may be necessary to obtain a more accurate transcript.

Robust-MSA then aligns the video with the transcript and offers customization of noise on a word-by-word basis. This enables researchers to simulate real-world scenarios where modality noise is present and evaluate how their models perform in such situations.

The platform visualizes the results of the videotext alignment for both the original and noise-injected versions of the video, highlighting the differences and potential impact on the model's predictions.

Noise Generation

In Multimodal Sentiment Analysis (MSA), modality noise can result in several common problems at the feature level. For example, facial detection failure may occur due to occlusion or a bad camera angle, leading to zero values in the corresponding feature dimension. Similarly, ineffective audio features may result from a noisy environment or bad microphone reception, while errors in automated speech recognition (ASR) algorithms or typos may introduce transcript errors and incorrect text features.

To help researchers simulate real-world data imperfections on a word-by-word basis, Robust-MSA provides six different noise simulation methods that can effectively mimic most modality noises encountered in real-world scenarios.

For video modality, the platform supports "Blank-Screen" and "GaussianBlur" methods, while for audio modality, it provides "Mute" and six different kinds of "additive background noise" from the DEMAND dataset [21]. For text modality, the options are "Replace" and "Remove." These methods can accurately simulate most modality noises from real-world scenarios, as they result in the same problems at the feature level.

In Robust-MSA, adding modality noise to a video is a straightforward process that can be accomplished with a simple drag-and-drop interface. As shown in Fig. 7.4a, researchers can simply select one of the six available noise simulation methods, drag it to a specific word in the transcript, and drop it onto the corresponding modality of the word. The platform will automatically apply the selected method to the corresponding modality, simulating the impact of modality noise on the model's performance. The added noise is highlighted with different background colors according to its modality, making it easy for researchers to identify and evaluate the impact of modality noise on their models.

Noise Defense Methods

In Robust-MSA, we offer three simple noise defense methods that researchers can use to improve the robustness of their models against modality noise. These defense

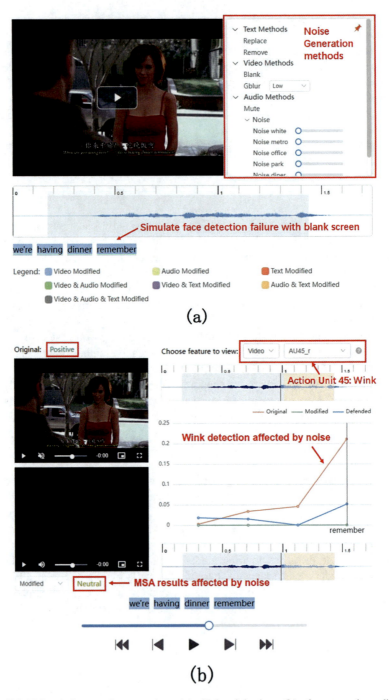

Fig. 7.4 Noise influence demonstration. (**a**): Noise injection, (**b**): feature and prediction comparison

methods include audio denoising, video motion compensated interpolation (MCI) at the raw data level, and feature interpolation at the feature level.

The audio denoising method is designed to denoise the raw audio wave using Fast Fourier Transform (FFT). The video MCI method generates missing frames using the Enhanced Predictive Zonal Search algorithm (EPZS) [22]. Finally, the feature interpolation method performs a linear interpolation on missing features.

After applying these defense methods, the defended video and features can be viewed on the final result page.

End-to-End MSA Pipeline

1. **Feature Extraction**: When it comes to real-time applications, it is essential to use the same modality features during both the training and inference stages of the MSA model. To achieve this, we adopted the eGeMapsv02 [8] feature set as acoustic features, while facial landmarks [10] and Action Units (AU) [12] were extracted as visual features. For textual features, we selected the BERT [14] language model.

   To enable video and audio to text alignment, we used a pretrained Wav2vec 2.0 model [9], which generates timestamps for alignment. All of these customized feature extraction processes were performed using the MMSA-FET toolkit [8], which helped us streamline the feature extraction process and ensure consistency across the training and inference stages of the model.

2. **Integrated MSA Models**: Robust-MSA currently supports eight state-of-the-art Multimodal Sentiment Analysis (MSA) benchmark models, including TFN [1], LMF [23], MISA [20], MAG-BERT [3], Self-MM [24], MMIM [25], and TFR-Net [26]. All of these models were trained on the MOSEI [6] dataset and have demonstrated high levels of accuracy and reliability in their predictions.

   To compare the performance of these models in noisy environments, Robust-MSA provides a platform where researchers can evaluate and compare the models' performance on real-world data with modality noise. The final sentiment prediction is shown in Fig. 7.4b, where Robust-MSA averages the models' outputs and maps the score into three classes: "Negative," "Neutral," and "Positive."

Noise Influence Demonstration

Robust-MSA provides researchers with a range of powerful visualization tools to help them better understand the impact of modality noise on extracted features and sentiment predictions in multimodal sentiment analysis.

To achieve this, Robust-MSA presents the original video alongside its noised-injected and noise-defended versions in alignment with the transcript. For each version of the video, the corresponding audio segment and text are highlighted, accompanied by the video. Researchers can click on a word or use the control buttons below to navigate quickly through words in the video and audio, enabling them to pinpoint segments where simulated modality noise is introduced with ease.

Moreover, Robust-MSA provides a line chart visualization of the modality features for all three versions of the video. This visualization allows researchers to compare the features across the three versions of the video and identify areas where feature interpolation or noise defense methods may be necessary. The x-axis of the chart represents corresponding words in the transcript, providing a clear and intuitive way to analyze the data, as shown in the right of Fig. 7.4.

7.2.3 Engaging the Audience

Our demonstration highlights the importance of model robustness in Multimodal Sentiment Analysis (MSA) tasks, even in the face of seemingly inconspicuous modality noise such as facial occlusion in a few frames.

To demonstrate this, we modified the original video by dropping the entire visual modality to simulate face detection failure, as shown in Fig. 7.4a. The results, shown in Fig. 7.4b, reveal that the noise-injected video is classified as "Neutral," while the original video is classified as "Positive."

By examining the features, we can clearly see that the absence of the "wink" action unit, which is a crucial visual cue for sentiment prediction, is responsible for the incorrect prediction.

7.2.4 Conclusion

In this section, we introduce Robust-MSA, a powerful visualization platform designed to help researchers gain a deeper understanding of the impact of modality noise on Multimodal Sentiment Analysis (MSA) models.

Robust-MSA offers an interactive visualization environment that enables researchers to explore the influence of modality noise on MSA models in real time. By presenting the original video alongside its noised-injected and noise-defended versions in alignment with the transcript, researchers can gain a better understanding of how modality noise affects the feature extraction process and leads to incorrect predictions.

In addition, Robust-MSA provides a timeline view of the modality features, enabling researchers to visualize how the features change over time and identify areas where modality noise may be impacting the model's performance.

7.3 Summary

This chapter introduces the M-SENA platform, an integrated tool for Multimodal Sentiment Analysis (MSA) researchers. It provides a comprehensive end-to-end solution for developing and evaluating MSA models, offering step-by-step instructions for data management, feature extraction, model training, and analysis. The platform includes evaluation tools, reliable benchmarks, and user-friendly visualization tools to aid researchers in understanding their models and identifying areas for improvement. Additionally, this chapter presents Robust-MSA, a powerful visualization platform that allows researchers to explore the impact of modality noise on MSA models in real time. By visualizing original and modified versions of videos alongside transcripts, researchers can gain insights into how modality noise affects feature extraction and predictions. The platform also provides a timeline view of modality features, enabling the identification of areas where modality noise may impact model performance.

References

1. Zadeh A, Chen M, Poria S, et al (2017) Tensor fusion network for multimodal sentiment analysis Proceedings of the 2017 Association for Computational Linguistics Conference on Empirical Methods in Natural Language Processing, 1103–1114
2. Zadeh A, Liang PP, Mazumder N, et al (2018) Memory fusion network for multi-view sequential learning. Proceedings of the 32nd Association for the Advancement of Artificial Intelligence Conference on Artificial Intelligence, 5634–5641
3. Wasifur Rahman, Hasan MK, Lee S, et al (2020) Integrating multimodal information in large pretrained transformers. Proceedings of the 58th Annual Meeting of the Association for Computational Linguistics, 2359–2369
4. Yu W, Xu H, Meng F, et al (2020) Ch-sims: A Chinese multimodal sentiment analysis dataset with fine-grained annotation of modality. Proceedings of the 58th Annual Meeting of the Association for Computational Linguistics. 3718–3727
5. Zadeh A, Zellers R, Pincus E, et al (2016) Mosi: multimodal corpus of sentiment intensity and subjectivity analysis in online opinion videos. arXiv preprint arXiv:1606.06259
6. Zadeh AAB, Liang PP, Poria S, et al (2018) Multimodal language analysis in the wild: Cmu-mosei dataset and interpretable dynamic fusion graph. Proceedings of the 56th Annual Meeting of the Association for Computational Linguistics, (1): 2236–2246
7. Björn Schuller, Stefan Steidl, Anton Batliner, et al (2016) The interspeech 2016 computational paralinguistics challenge: Deception, sincerity & native language. Proceedings of the 17th Annual Conference of the International Speech Communication Association, 2001–2005
8. Eyben F, Scherer KR, Schuller BW et al (2015) The Geneva minimalistic acoustic parameter set (GeMAPS) for voice research and affective computing. IEEE Trans Affect Comput 7(2): 190–202
9. Baevski A, Zhou H, Mohamed A, et al. (2020) wav2vec 2.0: A framework for self-supervised learning of speech representations. Proceedings of the 34th International Conference on Neural Information Processing Systems, 12449–12460
10. Zadeh A, Lim YC, Baltrusaitis T, et al. (2017) Convolutional experts constrained local model for 3d facial landmark detection. Proceedings of the Institute of Electrical and Electronics Engineers International Conference on Computer Vision Workshops, 2519–2528

11. Wood E, Baltrusaitis T, Zhang X, et al. (2015) Rendering of eyes for eye-shape registration and gaze estimation. Proceedings of the Institute of Electrical and Electronics Engineers International Conference on Computer Vision, 3756–3764
12. Baltrušaitis T, Mahmoud M, Robinson P (2015) Cross-dataset learning and person-speciffc normalisation for automatic action unit detection. Proceedings of the 11th Institute of Electrical and Electronics Engineers International Conference and Workshops on Automatic Face and Gesture Recognition, 6: 1–6
13. Pennington J, Socher R, Manning CD (2014) Glove: Global vectors for word representation. Proceedings of the 2014 conference on empirical methods in natural language processing, 1532–1543
14. Devlin J, Chang MW, Lee K, et al (2019) Bert: Pre-training of deep bidirectional transformers for language understanding. Proceedings of the 2019 Conference of the North American Chapter of the Association for Computational Linguistics: Human Language Technologies, (1):4171–4186
15. Liu Y, Ott M, Goyal N, et al (2021) Roberta: A robustly optimized bert pretraining approach. Proceedings of the 20th Chinese National Conference on Computational Linguistics, 1218–1227
16. El Ayadi M, Kamel MS, Karray F (2011) Survey on speech emotion recognition: features, classiffcation schemes, and databases. Pattern Recogn 44(3):572–587
17. Akçay MB, Oğuz K (2020) Speech emotion recognition: emotional models, databases, features, preprocessing methods, supporting modalities, and classifiers. Speech Comm 116:56–76
18. Qiu X, Sun T, Yige X et al (2020) Pre-trained models for natural language processing: a survey. Science China Technol Sci 63(10):1872–1897
19. Lian Z, Liu B, Tao J (2022) Smin: semi-supervised multi-modal interaction network for conversational emotion recognition. IEEE Trans Affect Comput
20. Hazarika D, Zimmermann R, Poria S (2020) Misa: Modality-invariant and-specific representations for multimodal sentiment analysis. Proceedings of the 28th Association for Computing Machinery International Conference on Multimedia, 1122–1131
21. Thiemann J, Ito N, Vincent E (2013) The diverse environments multi-channel acoustic noise database (demand): a database of multichannel environmental noise recordings[J]. J Acoust Soc Am 133(5):3591
22. Tourapis AM (2002) Enhanced predictive zonal search for single and multiple frame motion estimation[C]. Proc Vis Commun Image 4671:1069–1079
23. Liu Z, Shen Y, Lakshminarasimhan VB, et al (2018) Efficient low-rank multimodal fusion with modality-specific factors. Proceedings of the 56th Annual Meeting of the Association for Computational Linguistics, (1):2247–2256
24. Wenmeng Y, Hua X, Yuan Z et al (2021) Learning modality-specific representations with self-supervised multi-task learning for multimodal sentiment analysis. Proc Assoc Advanc Artific Intellig Conf Artific Intellig 35(12):10790–10797
25. Han W, Chen H, Poria S (2021) Improving multimodal fusion with hierarchical mutual information maximization for multimodal sentiment analysis. Proceedings of the 2021 Conference on Empirical Methods in Natural Language Processing, 9180–9192
26. Yuan Z, Li W, Xu H, et al (2021) Transformer based feature reconstruction network for robust multimodal sentiment analysis. Proceedings of the 29th Association for Computing Machinery International Conference on Multimedia, 4400–4407

Appendix

Symbol Cross-Reference Table

ASD	Active Speaker Detection
ASR	Automatic Speech Recognition
AV-MC	Acoustic Visual Mixup Consistent
BC-LSTM	Bidirectional Contextual Long Short-term Memory Networks
BERT	Bidirectional Encoder Representation from Transformers
CB	Character Bigram Features
CCAM	Channel Coattention Module
CHFFM	Contextual Heterogeneous Feature Fusion Model
CKSG	Character Based Key-substring-group Features
CNN	Convolutional Neural Network
CNP	Central Neural Processor
Corr	Pearson Correlation
CQT	Constant-Q Chromatogram
CS	Character Based Substring Features
CSFM	Contextual Single Feature Model
CSG	Character Based Substring-group Features
CSN	Cross-Stitch Network
CV	Computer Vision
DAE	Deep Automatic Encoders
DFG	Dynamic Fusion Graph
DNN	Deep Neural Networks
DP	Double Propagation
EF-LSTM	Early Fusion Long Short-term Memory Networks
EM	Expectation-Maximization Algorithm
FER	Facial Expression Recognition
FFT	Fast Fourier Transform
FLD	Facial Landmark Detection

(continued)

(continued)

Glove	Global Vector
GLUE	General Language Understanding Evaluation
GRU	Gate Recurrent Unit
HCI	Human-computer Interaction
HPS	Hard-Parameter Sharing
IAEE	Interaction of Aspect, Evaluation, and Emotion
IPAs	Intelligent Personal Assistants
LDA	Linear Discriminant Analysis
LF-DNN	Late-Fusion Deep Neural Networks
LLD	Low Level Descriptors
LMF	Low-rank Multimodal Fusion
LCS	Least Common Subsumer
LSTM	Long Short-term Memory Networks
MAE	Mean Absolute Error
MAG	Multimodal Adaptation Gate
MARN	Multiattention Recurrent Network
MCTN	Multimodal Cyclic Translation Network
MELD	Multimodal Emotion Lines Dataset
MFCC	Mel-Frequency Cepstral Coefficient
MFM	Multimodal Factorization Model
MFN	Memory Fusion Network
MLP	Multilayer Perceptron
MISA	Modality-Invariant and Modality-Specific Representations Network
MLMF	Multitask Low-rank Multimodal Fusion
MMIM	Multimodal InfoMax Network
MMix	Modality Mixup Strategy
mRMRs	maximum Correlation Minimum Eedundancy Algorithm
MSA	Multimodal Sentiment Analysis
M-SENA	a multifunctional and multimodal platform for displaying sentiment analysis
MTFN	Multitask Tensor Fusion Network
MTL	Multitask Learning
MU-SA	Multiutterance-Self Attention
MULT	Multimodal Transformer
NB	Naive Bayesian
NLP	Natural Language Processing
PCA	Principal Component Analysis
PS-MCNN	Partially Shared Multitask Convolutional Neural Network
RAF	Real-world Affective Faces
RCMSA	Residual Convolutional Model with Spatial Attention
RMFN	Recurrent Multistage Fusion Network
RNN	Recursive Neural Networks
SA	Sentiment Analysis
SC	Social Computing
SCAM	Spatial Coattention Module

(continued)

(continued)

SC-LSTM	Simple Contextual Long Short-term Memory Networks
SFEW	Static Facial Expressions in the Wild
SL	Soft-labeled Examples
STFT	Short-Term Fourier Transform
SVM	Support Vector Machine
TCDCN	Task-constrained Deep Convolutional Networks
TFN	Tensor Fusion Network
t-SNE	t-Distributed Stochastic Neighbor Embedding
WKSG	Word-Based Key-substring-group Features
WS	Word-Based Substring Features
WSG	Word-Based Substring-group Features

Code Link Table

CH-SIMS Dataset	https://github.com/thuiar/MMSA
CH-SIMS v2.0 Dataset	https://github.com/thuiar/ch-sims-v2
Co-attentive Multi-task Convolutional Neural Network	https://github.com/thuiar/cmcnn
HGFM: Hierarchical Grained and Feature Model	https://github.com/thuiar/HGFM
CM-BERT: Cross-Modal BERT	https://github.com/thuiar/Cross-Modal-BERT
Self-Supervised Multi-task Multimodal Sentiment Analysis Network	https://github.com/thuiar/Self-MM
TFR: Transformer-based Feature Reconstruction Network	https://github.com/thuiar/TFR-Net
M-SENA: An Integrated Platform for Multimodal Sentiment Analysis	https://github.com/thuiar/M-SENA

Printed in the United States
by Baker & Taylor Publisher Services